Bumblebees

Richard Comont

B L O O M S B U R Y
LONDON • NEW DELHI • NEW YORK • SYDNEY

BLOOMSBURY WILDLIFE
Bloomsbury Publishing Plc
50 Bedford Square, London, WC1B 3DP, UK

BLOOMSBURY, BLOOMSBURY WILDLIFE and the Diana logo are trademarks of
Bloomsbury Publishing Plc

First published in the United Kingdom 2017

A catalogue record for this book is available from the British Library

Library of Congress Cataloguing-in-Publication data has been applied for

ISBN: PB: 978-1-4729-6665-0; ePub: 978-1-4729-3363-8; ePDF: 978-1-4729-3362-1

4 6 8 10 9 7 5 3

Design by Susan McIntyre
Printed and bound in China by C&C Offset Printing Co., Ltd.

Published under licence from RSPB Sales Limited to raise awareness of the RSPB (charity registration in England and Wales no 207076 and Scotland no SC037654).

For all licensed products sold by Bloomsbury Publishing, Bloomsbury Publishing will donate a minimum of 2% from all sales to RSPB Sales Ltd, which gives all its distributable profits through Gift Aid to the RSPB.

Contents

Meet the Bumblebees

The low hum of bumblebees droning from flower to flower is one of the quintessential sounds of a British summer. The bees themselves are iconic, furry-bodied flying barrels. As well as helping produce the third of our food that is reliant on insect pollination, bumblebees have featured on postage stamps, given their names to film and TV characters, and even inspired a comic-book superhero.

Our familiarity with them is only skin-deep, however. The vast majority of the 'facts' attributed to bumblebees (that they dance, sting once and then die, swarm, or make honey) are actually true almost entirely of their thinner, balder domesticated relatives, the honeybees. The most famous myth of all – that bumblebees shouldn't be able to fly – is best disproved by watching one buzz busily between flowers. They're not graceful – 'to bumble' means to move awkwardly or ineptly – but they're definitely airborne, even when carrying half their own weight in pollen.

Most insects are seen as small, creeping or scuttling things, pests in the garden and a nuisance everywhere else. Bumblebees are different. With their bulky, furry bodies, low droning sound and habit of visiting flowers, bumblebees are some of Britain's favourite insects. The decline of bees in general is one of the most feared environmental problems and in the UK bumblebees even have an entire organisation – the Bumblebee Conservation Trust, with 10,000 members – devoted to monitoring and conserving them.

It comes as a surprise to many people that there's more than one species of bumblebee, but in fact there are around 250 species worldwide. Twenty-seven have at some point called Britain home, although nationwide extinctions have now left us with just 24 species. Worldwide, bumblebees are generally insects of temperate regions, and some species can be found well inside the Arctic Circle.
A few species do occur in the tropics, but most of those live in the cooler climate of the mountains.

Opposite: Queen Buff-tailed Bumblebee (*Bombus terrestris*) foraging on white clover.

Below: Bumblebees can remain active even at low temperatures.

Above: Early Bumblebee (*B. pratorum*) in a garden.

Because of their historical popularity with people, bumblebees have been given multiple common names. Only six species have just one known common name, and there are a total of 49 English names for the 27 British species. Even the group name has changed over time – 'bumblebee' was first recorded in 1530, 80 years after 'humblebee' was first given as a name for the group. In fact, humblebee was more common for centuries (Shakespeare and Darwin both mentioned humblebees), and the term only died out in the first half of the 20th century.

The genus name, *Bombus*, comes from the Latin for a loud buzzing, booming noise, while *Psithyrus*, the subgenus containing the quieter cuckoo bumblebees, means 'whispering'. Some species names show their close association with the land: *Bombus terrestris*, 'of the earth', for the hole-nesting Buff-tailed Bumblebee; while *B. pascuorum* and *B. pratorum* are 'of the pasture' and 'of the meadows', respectively.

Wanna-bees

Brightly coloured and distinctively patterned, bumblebees stand out from their background. Armed with a venomous sting, they use warning colouration rather than camouflage to avoid predation, trusting that most of their natural enemies have learned that striped black and yellow spells danger. The fewer patterns that a predator has to learn, the better for prey – which is why so many bumblebees look very similar to each other.

Above: Bee Chafer *Trichius fasciatus*, a beetle which mimics bumblebees.

If imitation is the sincerest form of flattery, then bumblebees' fan club extends well beyond humankind. Several other insects, including hoverflies, bee flies and the Maid of Kent rove beetle (*Emus hirtus*), are good mimics of bumblebees, copying their fur, their colouration and even their flight patterns. Some mimics, such as the hoverflies *Merodon equestris* and *Volucella bombylans*, even have several colour forms to mimic different species of bumblebee.

All this mimicry is done for a reason. It's a hard life being an insect, with so many bigger things keen to eat you, and a bumblebee's sting is one of the few effective weapons against these threats. If a mimic can make a predator think – even for a millisecond – that it needs to be a bit careful when attacking, this gives the prey extra time to get away.

Right: The bee-mimic hoverfly *Volucella bombylans* – this time impersonating the Common Carder Bumblebee (*B. pascuorum*).

Feeding anatomy

Above: Bumblebees are covered from head to tail with branched, almost feather-like hairs - its pollen-collecting kit.

Bumblebees, like all insects, wear their skeleton on the outside. This exoskeleton is made of hard plates of chitin, joined together by flexible sections, and, unlike the hair covering it, is completely black. Like humans, bumblebees lose their hair as they get older (mainly rubbing it off against flowers), and late-summer bumblebees can be almost bald and completely black, making identification very difficult.

Bumblebees are very well adapted to collect pollen and nectar as efficiently as possible from flowers, and their hair has a role to play here as well. The hairs that cover a bumblebee's body are branched, and, under a microscope, look almost like sparse feathers. These help generate a static charge as the bee flies, attracting pollen grains as the bee lands on a flower and encouraging them to stick to the hairs. Bees then groom themselves, using the array of spines and stiff bristles on their legs to comb the pollen out of their fur. This action also gradually pushes the pollen into the pollen baskets on the tibia (shin) of female bumblebees' hind legs. This is a flattened, smooth area, fringed with long hairs that act as scaffolding around the outside. Pollen, wetted with a little nectar, builds up in a lump with the consistency of plasticine. A full pollen basket may contain as many as a million individual grains of pollen.

Above: A Garden bumblebee with half-full pollen baskets.

A flower's nectar is generally stored deep inside and needs a specialised tool to be accessed. This means a tongue up to 20mm (¾in) long, which is hairy at the end in order to act as a brush and soak up the nectar from flowers. The bee repeatedly dips its tongue into the nectar, and the fluid soaks its way up the tongue and into the bee's mouth – they don't suck it up! Surrounded by a four-part sheath, the tongue is held under the bee's body when not in use, but is held out in front of the bee like a probe as it approaches flowers. The nectar is stored in a special honey stomach (a large bag at the front of the bee's digestive system), which when full – perhaps only after visiting hundreds of flowers – may fill 95 per cent of the abdomen and account for 90 per cent of the total body weight of the bee.

Left: A Heath Bumblebee (*B. jonellus*) nectaring, its long tongue inserted into the flower.

General anatomy

Bumblebees, like all insects, have their bodies divided into three main parts: the head, thorax and abdomen. As well as the tongue, the head also contains the jaws. In bumblebees, these are substantial, blunt-ended and sideways-biting: the bee uses them mainly to eat pollen.

On the top of the head, bumblebees carry their sensory apparatus. Each bee has two antennae, which sense chemicals in the air, two large compound eyes, and three small primitive eyes (ocelli). The head doesn't carry ears, though – bumblebees don't have ears and can't hear, though they can sense vibrations, for instance through whatever they're standing on.

The compound eyes, with their hexagonal lenses, are the main way that the bees see the world – and they see it very differently to us. Bumblebees can't see many colours towards the warmer end of the spectrum as their eyes do not contain red-sensitive cells; instead, they have cells

Below: Face to face with a bumblebee: big eyes, big mandibles.

Basic bumblebee anatomy

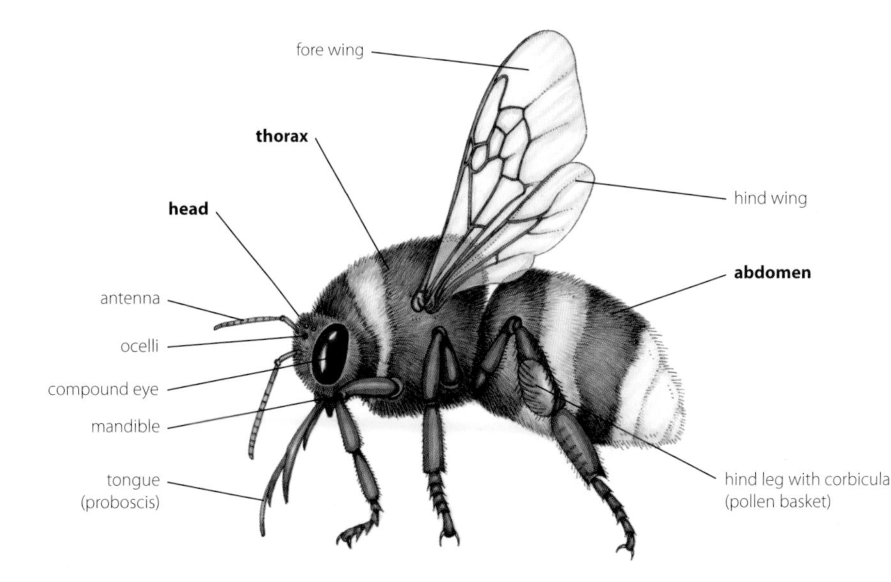

which can detect green, blue and ultraviolet colours, which enables them to see patterns on flowers that are invisible to humans. The three ocelli look like small, shiny domes, and are arranged in a triangle on top of the head between the big compound eyes. They detect light intensity and help the bumblebee tell which way is up, and what time of day it is.

The abdomen contains the honey stomach, the digestive system and the reproductive system. In queens and workers it also contains the sting, which is a modified part of the reproductive system, and the venom system which supplies it. Bumblebees don't die as a result of stinging someone or something – they have a smooth sting. Honeybees have barbed stings which get stuck in the wound they create by stinging, which tears the venom system out of the bee as it tries to escape, killing the bee. But bumblebees can simply pull out their non-barbed sting and make their escape.

Below: The barbed sting of a honeybee.

Wings and things

Above: Worker Red-tailed Bumblebee (*B. lapidarius*) flying between flowers.

The midsection of a bumblebee – the thorax – is the bee's power station, chock-full of muscles. The huge wing muscles take up much of the space, propelling the bumblebee into the sky at more than 8,000 wingbeats per minute. Unusually, the muscles must actually vibrate like an elastic band to flap the wings this fast, as the nerves controlling the wing muscles can't fire this frequently. The bumblebee's buzz comes from this process but not from the wings themselves – instead, it's from the air rushing through the bee's breathing pores (spiracles) to supply oxygen to the wing muscles.

Flapping this fast also generates a huge amount of heat – so much so that bumblebees are essentially warm-blooded creatures. For a bee to be able to fly, the internal temperature of the thorax has to be at least 30°C (86°F), and it achieves this by disconnecting its wing muscles from the wings and vibrating them – essentially, by shivering. This means bumblebees can fly in cooler weather than most other insects, visiting flowers early in spring, late in autumn, and in the mornings and evenings when few other pollinators are out and about. It can take a few minutes for the thorax to heat up sufficiently – around five minutes from a resting temperature of 13°C (55°F) – and the temperature is maintained between 30 and 40°C (86 and 104°F) during flight, regardless of the air temperature.

The bee flight myth

One of the best-known things about bumblebees is that 'scientists have proved' that they can't fly. As they so evidently can and do fly, the tale is often used as evidence that scientists have no idea what happens in the real world. Alternatively, the story is sometimes told as support for the power of self-belief, often with the addition that if the bumblebees were told they couldn't fly, they would no longer be able to, as if the bees were like Wile E. Coyote, running off the edge of a cliff but not succumbing to gravity until a downwards glance revealed nothing more substantial than air beneath his feet.

The myth seems to have started in the first half of the 20th century, when a physicist tried to apply the new science of aerodynamics to a bumblebee. A few calculations later and he discovered, presumably to his shock, that the wings of his bumblebee were too small to keep the bee airborne, and an urban legend was born.

What the physicist had actually shown was that bumblebees could not fly by simply holding their wings out straight and hoping for the best, like an aeroplane. This is entirely true, and the glide path of a bumblebee tends to be short, unceremonious and vertical – you can test it yourself the next time you find a dead bumblebee, by dropping it and seeing how far away it ends up.

Above: A helicopter has small wings (rotor blades) but they rotate very fast to generate lift.

Instead, bumblebees' flight is less like that of a plane and much more like that of a helicopter. Bumblebees don't simply flap their wings up and down; instead, a complex flapping motion sees the wings twisted through the air, with the wing-tips actually performing a figure-eight motion for each stroke. At up to 200 wingbeats per second, the bending and flexing of the wings and the complexity of the stroke combine to generate the extra lift a bumblebee needs to stay in the air, and it flies on.

Bumblebees do this by balancing how much blood they pump to the colder abdomen, where it cools down – on hot days, the bumblebees may fly with their abdomens fully extended and full of blood to stay cool, while on cold days the blood may stay largely in the warm thorax.

This all comes at a cost, however. The huge wing muscles need a big supply of energy to work – the metabolic rate of a flying bumblebee can be up to 75 per cent more than that of a hummingbird. Consequently, a bumblebee is never more than about 40 minutes' flight away from the fuel exhaustion point – it needs to keep filling up with high-energy nectar from flowers.

Relatives and Predecessors

Bumblebees are members of the family Apidae, the largest bee family worldwide. What exactly makes a bumblebee – and how bumblebee species are related to each other, and to other bees – can be complex and contradictory. This is made harder by the relative lack of bumblebee fossils, and the scale of the identifying features: even today, with live specimens, identification of species can need a microscope. To understand what makes a bumblebee a bumblebee, we must go back in time 135 million years, to the evolution of flowering plants, before scrolling forwards to watch that first tentative four-winged flower visitor become the friendly flying teddy bear of today.

The bee family

Today's bumblebees are all within the same genus, *Bombus*, which makes all the species sisters, as closely related to each other as dogs are to wolves and coyotes. The genus *Bombus* makes up a small part of the family Apidae, which contains 20,000–30,000 bee species worldwide. Together with ants and wasps, bees make up the group Aculeata (from the Latin *aculeus*, 'sting'). These join parasitic wasps, gall wasps and sawflies to make the order Hymenoptera ('membrane-winged' or 'married-winged' from either the ancient Greek *hymen*, a membrane, or Hymen, the ancient Greek goddess of marriage, after the row of hooks that 'marry' the wings together in flight). With around 150,000 known species, Hymenoptera is the third-largest insect order, behind beetles and flies, and is the second-biggest UK insect order with around 6,000 British species.

Above: A Gall wasp – a non-bee member of the family Hymenoptera.

In Britain, 27 different bumblebee species have been recorded. Of these, the Apple Bumblebee (*B. pomorum*) was only ever found at Deal and Kingsdown, Kent, in

Opposite: A bumblebee and a honeybee share a flower.

the mid-1850s and 1860s, where it was almost certainly a natural colonist that survived for a few years before dying out. Two species went extinct in the UK during the 20th century: Cullum's Bumblebee (*B. cullumanus*), last recorded near Blewbury in the Berkshire Downs in 1941, and the Short-haired Bumblebee (*B. subterraneus*), last recorded as a British species in 1988 at Dungeness, Kent. In 2009, work began to reintroduce the latter species from Sweden, but has yet to be confirmed as a British breeding species. Cullum's Bumblebee, however, appears to be gone for good. Always a rarity in Britain, it seems to have survived solely on the extensive chalk grasslands of southern England and as these were broken up, the bees died out – a sad end for a species first known to science from a corner of Suffolk, but now extinct across the country which discovered it. With these three losses, Britain currently has 24 established bumblebee species.

Below: Specimens in museums are all that remain of the British population of Cullum's Bumblebee (*B. cullumanus*).

The extended family: bees beyond bumbles

Above: A domesticated Western Honeybee.

Honeybees

Of course, there's more to bees than just bumblebees. Probably the most famous bee – and possibly the best-known insect – is the Western Honeybee (*Apis mellifera*), and the majority of things that most people know about bees actually just apply to this unusual and highly social insect.

One of very few semi-domesticated insects, this thin and less hairy bee's history has been intertwined with that of humankind for millennia as we have used them for honey, beeswax and crop pollination. Rock paintings in Spain dating from around 8000 BC show a honey hunt up a cliff; chemical analysis of prehistoric pottery has found beeswax in use across Europe from at least 6500 BC. The ancient Egyptians kept honeybees: one of the Pharaoh's titles was 'King of the Bees', and the earliest pictures of beekeeping are in temples and tombs from 2400 BC. In fact, the earliest known honey sample

Below: Bees depicted in an ancient Egyptian carving.

Above: Keeping honeybees in hives means the bees can be moved, ensuring a supply of honey when and where required.

was found in an Egyptian tomb and dates to around 1000 BC – and was (supposedly) still edible when discovered, 3,000 years after being stored in the darkness of the tomb.

In the UK, the vast majority of the honeybees you see will be domesticated, foraging from someone's hive. Feral hives do exist, generally setting up home in tree cavities after swarming from a beekeeper's hive, but the parade of parasites and diseases that have affected beekeeping around the world for the past century, from Isle of Wight disease to the Varroa Mite (*Varroa destructor*), have taken a huge toll on these wild populations.

Left: A wild honeybee nest extending out of a tree cavity.

Solitary bees

Bumblebees and honeybees, with 250 and 5 species worldwide respectively, are massively outnumbered by the solitary bees. These are hugely diverse as a group: around 225 of the UK's 250 bee species are solitary, and around 680 of Europe's 750. The term 'solitary bees' comes from their solitary nesting arrangements (each female builds her own nest and there are no workers), but often several females will build nests together – in a garden bee house, for example.

Unsurprisingly, given the number of species, solitary bees have evolved to fill a wide range of niches. The majority have what might be called 'normal' bee lifestyles – they hatch from the nest, feed from flowers, mate, build nests of their own, and stock them with nectar and pollen to sustain their own offspring. The form and placement of the nests varies hugely – leafcutter bees in the genus *Megachile* cut neat chunks from leaves to build brood

Below: The Ivy Bee (*Colletes hederae*), returning to its sandy nest burrow.

Above: A Red Mason Bee filling mud cells with pollen in a bee hotel.

chambers in cracks and crevices, while the Red Mason Bee (*Osmia bicornis*) does the same thing with mud. Most of the 68 species in the genus *Andrena* dig burrows in soil, excavating side-chambers which they fill with pollen; *Colletes* species do the same but store more nectar and live in sandier soil, so they line their brood cells with a glue that they secrete themselves, which dries like a waterproof cellophane wrapper.

My personal favourite is the relatively scarce Red-tailed Mining Bee (*Osmia bicolor*), a small, furry, bright red and black bee which lives mostly in good-quality chalk or limestone grassland in the south of England. This species doesn't bother with burrows – instead, it seeks out old snail shells, from big species like the banded snails. Once she's found a shell, the female cleans it out, then fills it with a mixture of pollen and nectar, and lays a single egg. She then wrestles the shell into a mouth-down position, before beginning her second bout of foraging – this time for pine needles, dry grass and small sticks, which she clutches with her mandibles as she flies back to the nest – a bee riding a broomstick! – before stacking them neatly on and around the shell. No one knows for certain why she does this – it's generally assumed to be camouflage against predators – but it's an amazing thing to watch.

Above: Red-tailed Mining Bee arranging dead grass stems around its snail-shell nest.

There are also plenty of species whose ancestors clearly thought that foraging and nest-building was a bit too much like hard work. Bees in at least six genera (*Nomada, Melecta, Stelis, Sphecodes, Epeolus* and *Coelioxys*) are parasitic on the various nest-making species. They do no foraging themselves and simply hang around the nests of the foragers, waiting for the opportunity to sneak into the nest and lay their own egg into the brood chamber, often killing the original egg in the process.

Below: A Nomad Bee (*Nomada flava*) nectaring on forget-me-not.

The vegetarian wasps

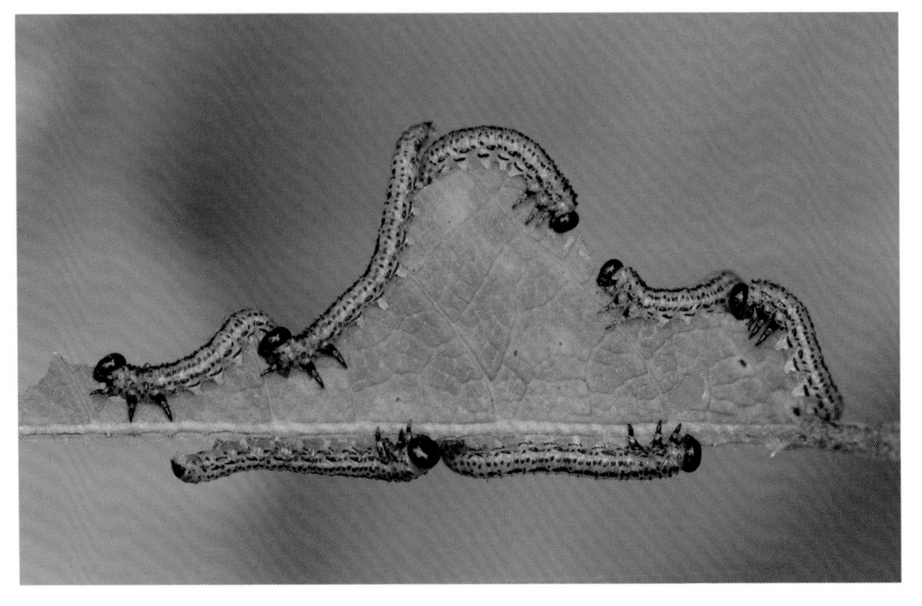

Above: Caterpillar-like Birch Sawfly (*Craesus septentrionalis*) larvae eat leaves, not pollen.

The first fossil Hymenoptera known to science appear during the middle to late Triassic period (245–208 million years ago), in what is now Kyrgyzstan. Rather than the specialised bees of today, these were sawflies, primitive wasp-like creatures with caterpillar-like, plant-eating larvae and no pinched-in wasp 'waist'. Ants and wasps, close relatives of the bees, began to appear during the late Jurassic period. Unlike the herbivorous sawflies, these were at least partially predatory, gathering their protein by hunting other insects.

But meat is not the only protein source, as any vegetarian can tell you. Plant pollen also contains a high proportion of protein, up to 40 per cent in some plant species, and by this time had already existed for millennia, produced in huge quantities by the conifers that almost monopolised the Jurassic

Below: A thick-waisted adult Birch Sawfly similar to an ancestral bee.

forests. Around 135 million years ago, in the early Cretaceous period, plants began to take the next step forward: the evolution of flowers. For the first time, pollen was packaged in a way that attracted animals, rather than relying on the wind to disperse it to just the right spot at the right time for pollination to happen. The starting gun had been fired on the race to exploit the new resource.

Above: A spider-hunting wasp with its spider prey.

The first flowers were simple bowls and the first pollinators probably beetles – some plants, such as magnolias, are still almost entirely beetle-pollinated today – but it didn't take long, in evolutionary terms, for the first bee to arrive on the scene. The very first bees evolved from an insect-hunting wasp in the family Crabronidae around 125 million years ago, during the early Cretaceous period. As the wasps discovered that pollen was an easier protein source than hunting down other insects, they became gradually better adapted to collecting it from flowers. This triggered a huge, fast starburst of species creation. As both bees and plants started to become specialised, they pulled away from the generic ancestral bee or plant and split off from each other as separate species. The mutual relationship between bees and flowers, which has become fundamental to life on Earth, had begun.

Right: Magnolia flowers like this one are still pollinated by beetles.

Bumblebees arrive on the scene

These primitive Cretaceous bees were still a long way from being bumblebees. The very first fossil bee known to science, *Melittosphex burmensis*, dates from around 100 million years ago and was found preserved in amber in Myanmar. Unsurprisingly, it has many primitive features, including an absence of the familiar pollen baskets of today's bumblebees and honeybees. These structures are thought to have evolved around 80–90 million years ago. However, the first fossil evidence of them comes in the form of a piece of amber from New Jersey, USA, containing the bee *Cretotrigona prisca* from shortly before the dinosaur-eradicating K-T extinction event, 65 million years ago. Despite its age, *C. prisca* is very similar to the modern-day stingless bees of Central and South America. These, like bumblebees and honeybees, are highly evolved social species, but are cousins, not ancestors, of modern-day bumblebees.

There are a handful of fossil bumblebee-ancestors dotted around the world's museums today. Almost all are of individuals which died in lakes or mudflats and were entombed between the layers for millennia. The first species which are recognisably closely related to today's bumblebees and honeybees are *Calyptapis florissantensis* and *Oligobombus cuspidatus*, both dating from 33.9 to 37.2 million years ago and found in Missouri, USA, and the Isle of Wight, UK, respectively. These species were flying on the cusp of the emergence of the true bumblebees, but we aren't quite there yet. For them, we need to be in a different continent entirely.

We need to go, in fact, to Asia, to the region that is now the Himalayas. Thirty-four million years ago,

Below: South American stingless bees perched on their nest funnel.

the mountains themselves weren't yet there, but there was a cooling event in the regional climate. The local bees responded to the colder weather by becoming larger, rounder and hairier, in order to minimise heat loss. As the Himalayas rose, these big furry proto-bumblebees continued to adapt to the conditions, diversifying into several species – the area around Tibet and south-western China remains a bumblebee hotspot even today.

These cold-adapted bumblebees couldn't disperse south, into the Indian jungle – the same large size and hairiness that preserved them through the Himalayan snow meant they couldn't survive in the tropics without overheating. The north was cooler, and so the bumblebee species dispersed northwards into the temperate zones, reaching Europe and North America. The bees adapted to the new conditions and were soon noticeably different from those which had remained in the Himalayan region.

Above: Bumblebees evolved in mountainous areas and are still adapted to live in cooler climates.

Around the world, fossils of almost a dozen species of bumblebee have been found in ancient mudflats, dating from between 5 and 20 million years ago. All are similar enough to today's bumblebees to be placed in the same genus, *Bombus*. One is from North America – *B. proavus*, fossilised between 11 and 20 million years ago in Washington – and a handful have been found in China (*B. luianus*, *B. dilectus* and *B. anacolus*, all dating to 16–12 million years ago), but the majority so far have turned up in Europe, from Greece in the south to Russia in the north. These temperate regions are still heartlands for bumblebees today; France and the USA both have 46 species, and even Britain's 27 recorded species are almost 11 per cent of the global total.

Bumblebees worldwide

Above: Bumblebees are less common in the tropics, but some species like this White-tailed Bumblebee can survive hotter climates.

Although bumblebees are most numerous in the temperate regions of Europe, Asia and North America, they can be found almost everywhere worldwide. In the north, some species' ranges extend right into the Arctic Circle, while the Patagonian Bumblebee or 'flying mouse' (*B. dahlbomii*) can be found as far south as Tierra del Fuego at the tip of South America. Australasia is the only continent without native bumblebees and even here three species can now be found, introduced to pollinate crops at the end of the 19th century.

Arctic bumblebees

Taking the cold-hardiness of their ancestors to extremes, two species commonly live inside the Arctic Circle, less than a thousand kilometres (620 miles) from the North Pole. These are the Polar Bumblebee (*B. polaris*) and the High Arctic Bumblebee (*B. hyperboreus*), and they're the northernmost bumblebee species in the world. The Polar Bumblebee is the more common of the two; thickly covered with long black and yellow hair, it can be a frequent sight in the spring and early summer on the

Arctic tundra, buzzing between the flowers of lousewort and other hardy plants.

Above: The High Arctic Bumblebee.

Unlike many of the other large insects found this far north, which can take many years to reach maturity (one, the Isabelline Tiger Moth (*Pyrrharctia isabella*), can spend 12 years as a caterpillar and freezes solid each winter), the bumblebees must emerge from hibernation, build their nests, raise a worker brood and then produce males and new queens in the eyeblink of an Arctic summer. Even with the bees' special abilities – they can raise their body temperature to 38°C (100°F), warmer than humans, and use the heat to begin incubating their eggs before they're even laid – there's only a month or two a year when activity is possible. The only way the species can survive is to cut the colony-building phase to the bone. For the Polar Bumblebee,

Below: Caterpillar of the Isabelline Tiger Moth.

this can often mean rearing just the one batch of workers in the heavily insulated nest before switching to raising males and new queens.

For the High Arctic Bumblebee, things take a more sinister twist. While queens can – and sometimes do – build their own nests and rear their own workers, more often they seek out new nests of the closely related Polar Bumblebee and take them over, laying their own eggs and using the workers as a slave-labour force to produce males and new queens.

South American bumblebees

In the south, there are no bumblebees in Antarctica, but there is one native species in southern South America, and it's found right down into the wintry landscapes of Tierra del Fuego. This is the Patagonian Bumblebee, which is a brilliant rich ginger all over and the largest bumblebee species in the world, with queens reaching 4cm (1½in) long. Unusually for bumblebees, this species is a regular visitor to red flowers, including the national flower of Chile, Chilean Bellflower, which is otherwise pollinated by hummingbirds.

Below: Earth's largest bumblebee, the Patagonian Bumblebee, a species sadly in decline.

Unfortunately, the scarcity of bumblebees in temperate South America led to two European species being introduced as crop pollinators. The Ruderal Bumblebee (*B. ruderatus*) and the Buff-tailed Bumblebee (*B. terrestris*) were introduced into Chile in 1982 and 1998 respectively, and populations of the native species have declined ever since. The invading species both outcompete the Patagonian Bumblebee (by flying earlier in the year, visiting more flower species and producing more new queens) and also wage biological warfare. The Buff-tailed Bumblebee in particular carries the parasite *Apicystis bombi,* which interferes with foraging and colony formation. A relatively minor problem for the co-evolved Buff-tailed Bumblebee, the parasite has a devastating effect on the undefended native species, which is now classed as 'Endangered' on the IUCN Red List.

There are some bumblebees which survive and thrive in warmer regions, including several in the northern half of South America. One of these – the Black Bumblebee (*B. atratus*) – is unique as the only bumblebee species which can have several active queens in a nest at once. Living in the tropics, there is no need for hibernation and the species often has nests which can last several years – more like the honeybees than most other bumblebee species. One queen (the founder) builds a nest, as usual, but when she dies, instead of the colony remnants dispersing, she can be replaced by one or more new queens. These can be 'true' queens, raised for the purpose, or 'false queens', large workers who mate with a male and begin to lay both male and female eggs.

Below: Chilean Bellflower, pollinated by hummingbirds and the Patagonian Bumblebee.

Sometimes the queens can coexist and the colony continues with two or even three queens together; often, a queen will become dominant by killing or subduing her rivals for the throne. When this happens the colony will revert to a more normal single-queen arrangement, and subordinate queens will either leave or take on the role of a worker. Colonies usually switch between these one-queen and multiple-queen phases several times per year.

What Makes a Bumblebee?

As a bumblebee colony winds down at the end of summer, the new queens are produced. Each of them has the chance to establish a new colony – if they get that far. First they need to find a mate, then they need to survive the winter and then find a nest site of their own. This won't be easy...

Finding a mate

Opposite: Mating pair of bumblebees, male on top.

Queens are the largest individuals of any species of bumblebee, and need a lot of food growing up. They're only produced towards the end of a colony's existence, once the workers have collected a decent amount of stored pollen. For most species in the UK this happens in August or September, and queens are produced a week or so after males from the same nest, which minimises the chances of inbreeding.

The males often look very different from the workers and queens. Males don't have pollen baskets – instead their hind legs are hairy and rounded rather than flattened. They have much more hair on their faces than do females, and their hair all over tends to be longer, shaggier and more unkempt-looking than that of either queens or workers. Males also have a distinct tendency to have a lot more yellow on them than females of the same species have – so if you see a blonde bumblebee with hairy legs, a big moustache and bed hair, it's likely to be a male!

In British bumblebees, the most extreme example of this difference between the sexes is the Red-tailed Bumblebee, *Bombus lapidarius*. Females – both queens and workers – are jet black all over apart from their rich red tail. By contrast, as well as having the red tail, males also have a thick band of bright yellow hair at the very front of the thorax that extends forwards onto the face, and a sparser yellow band of hair at the back of the thorax.

Below: A male Early Bumblebee (*B. pratorum*), complete with yellow moustache.

Mating

Above: With no foraging to do, male bumblebees can often be found sitting on leaves or flowers.

Males also behave differently. Freed of the burden of foraging, there's no need to buzz busily between flowers. Instead, they can devote their time to finding a newly emerged queen in order to mate with her. Different species have different approaches to this. Some – most notably the Tree Bumblebee (*B. hypnorum*) – gather outside a nest, a swirling maelstrom of males waiting for a new queen to run the gauntlet. When she does, males fly at her from all directions, trying to grab her, and they fall to earth. The most persistent male mates with her on the ground, while the rest take off and resume patrolling the nest entrance, churning like tea leaves in a cup.

Males of some species – particularly the Nevada Bumblebee (*B. nevadensis*) and the Tibetan species *B. rufofasciatus* – are territorial. They have much bigger eyes than most other species, an adaptation to see movement better in their chosen patch. They spend their time perched on a prominent vantage point like a twig, and zip out to chase away any other males or intercept passing queens.

Males of most species spend their time patrolling a regular circuit, looking out for queens. In Britain, these circuits are usually 100–1,000m (328–3,280ft) long and have around 30 or so males flying round at any one time. The largest known, in Central America, can be 2.5km (1½ miles) long and contain up to 720 males of *B. pullatus*. Interference between species is minimised because different species reach peak activity at different times of the day. Also, they fly circuits at different heights:

Above: A cloud of male Tree Bumblebees waiting for a queen to emerge from a nest inside the birdbox.

Buff-tailed and Garden Bumblebees (*B. terrestris* and *B. hortorum*) fly less than a metre (3¼ft) from the ground, White-tailed at 5–10m (16½–33ft), and Red-tailed up to 17m (56ft) high, up in the treetops. Some, such as the Red-shanked Carder Bumblebee (*B. rudararius*), will fly in small groups, but most species spread out to limit the competition between individuals.

Males don't just fly round these racecourse circuits – they also use pheromones. Different species use different cocktails of scents, but many are lemony and most can be detected by people. To me, male Early Bumblebees smell like citronella, while Buff-tailed Bumblebees smell more musty. On their circuits, the bees mark plant stems by running up and down them, repeatedly chomping their

Above: Some male bumblebees, like this Common Carder Bumblebee (*B. pascuorum*), look very similar to females of the same species but they don't have pollen baskets.

mandibles on the stem. This releases the pheromones from glands on the outside of each mandible, and the male's facial hair acts as a paintbrush, daubing the chemicals onto the marker.

New queens coming across these scent blobs will land nearby and wait for a male to arrive, then mate with him. Queens more than 40cm (15¾in) from the scent markers are not usually noticed by the males – even when the two are feeding from the same flower. The circuit itself tends to be fairly poor for flowers, so males must take breaks from their patrolling to divert to nearby flowers for refuelling. These are likely to be frequented by many other species of bumblebees – males, workers and queens – so keeping segregation of mating areas and feeding areas ensures the minimum of wasted effort attempting to mate with the wrong species or sex.

One nest or two?

Most of the time, males and new queens will emerge during August/September, and mated queens will overwinter and produce new nests in the spring. A reasonable proportion of bumblebees, however, will have two generations a year. For these, the queens produced from the first nests will mate and produce their own nests without an intervening overwintering stage. It's especially common in early-emerging species like the Early and Tree Bumblebees, but has been recorded in many other species, particularly in years with warm springs and long summers.

The Buff-tailed Bumblebee has taken things one step further, and a proportion of the population is now frequently active all year round in the warmer parts of Britain. The combination of generally warmer winters (particularly in cities) and increased planting of winter-flowering plants has allowed the species to establish nests in October/November. Workers forage largely on planted mahonia, crocuses and winter-flowering heathers, with males and new queens thought to be produced in February–March as the queens, which instead decided to hibernate, come out of their winter hideaways.

Above: The more pollen bumblebees bring back to the nest, the better.

Below: Increased planting of winter flowers has allowed some bumblebee species to fly all winter.

Hibernation

Above: Bumblebees emerging too early in the year can be caught out – and sometimes killed – by low temperatures or sudden snowfalls. Antifreeze-filled blood is not enough without shelter.

Only new queen bumblebees hibernate. All others - males, workers, old queens - will die before winter comes. Before hibernation, new queens mostly gorge on the food brought back by their sister workers, away from the threat of predators outside the nest. Once they leave they'll still visit flowers, but only to feed themselves – they won't collect pollen until they have a nest of their own. Once they've reached Christmas-lunch levels of fullness, it's time to find a place to spend the winter.

Perhaps counter-intuitively, queens look for somewhere relatively cold to wait out the winter. Cool, dry places tend to maximise survival over the six months or so that the queen will sleep. Wet places will encourage fungus growth – the soil she's resting in contains the spores of many species of fungus specialised in growing on and killing insects, waiting for warm, wet conditions and a suitable host. And sites that are too warm make the queen wake up too early in the year. This either means she leaves her overwintering site just in time to be caught by a late wintry blast, or just that she burns through her food stores too quickly and runs out before spring.

Some queens hollow out a snug cavity in rotten logs, beneath stones, or underneath thick moss blankets at the base of trees. Others tunnel into the cool north side of a bank, using their powerful jaws and front legs to dig and reversing back out of the hole to bulldoze loose debris out of the shaft. After one or two hours of digging, the queen will have excavated a short (5–15cm/2–6in) burrow. It's a delicate balancing act – the deeper she goes, the safer she'll be from the weather, but the more energy she will use. Once safely tucked up, her metabolism slows down and she enters a deep sleep called torpor to minimise her energy use. She will stay like this for anywhere from five to 11 months, before warmer temperatures signal that winter is in retreat.

Even if they find the ideal cool, dry overwintering site, around half of the queens that enter dormancy in autumn don't make it through to spring. Surprisingly few die of cold, at least in Britain; most British species produce antifreeze in their blood over the winter and can survive temperatures well below zero for a reasonable length of time. Some simply don't have enough fat on board to make it through the winter, and die in their sleep; others succumb to natural disasters (winter flooding, for example). Yet more are eaten by hungry mice and other predators, or die of diseases or parasites contracted from the soil or even during the summer.

Below: Before hibernation, new queens mostly feed inside the nest to minimise the risk of predation.

Emergence at last

Above: Bumblebees can often be found using spring flowers such as daffodils and bluebells.

Below: Sallow catkins (pussy willows) are a fantastic food source for hungry queens.

The first bumblebee of the year is always a true sign spring is on the way. In the UK, the earliest species to emerge is usually the Buff-tailed Bumblebee in February or early March, and the Early and Tree Bumblebees complete the trio of early risers. They can be seen visiting those other harbingers of spring, bluebells and daffodils. The queen bumblebee's first nourishment of the new year, however, is actually her final taste of last summer. Often before even leaving her overwintering site, she will drink the contents of her honey stomach filled the previous summer from the communal honeypots in her birth nest.

She will spend some time – often a fortnight or more – feeding herself up and replacing the fat she lost during the winter. Flowers tend to be thin on the ground at this time of year, and the nectar from early-flowering plants, such as pussy willows, are vital for feeding up the undernourished queen. She must be well fed for her ovaries to mature in order for her to lay her eggs and begin the cycle again.

Dispersal

Above: Once refreshed, queen bumblebees disperse looking for nest sites.

First, though, she must find her own place to nest. Some will disperse long distances, often at relatively high altitudes, before looking for nest sites; studies have found that the Red-tailed and Common Carder Bumblebees frequently disperse 3-5km (1¾ –3 miles) from their birth nest, and several species have been found many kilometres offshore on ferries, lightships and other vessels. Queens of the White-tailed Bumblebee (*B. lucorum*) have even been seen in flight in the middle of the 80km (50 mile) wide Gulf of Finland! These movements can involve huge numbers as well: there are Scandinavian records of 900 queen White-tailed Bumblebees passing through a 150m (490ft) coastal strip in an hour.

Most queens don't disperse very far at all. Different subspecies of the Buff-tailed Bumblebee (indicating well-separated populations) are found on either side of small sea crossings – between the UK and Europe (32km/20 miles), Spain and North Africa (16km/10 miles), Italy and Sardinia (12km/7½ miles), and Corsica and Elbe (10km/6¼ miles).

A nest of her own

Above: Many species prefer to nest in a hole, often an old mouse burrow.

When she is in an area she likes the looks of, a queen begins to show another new behaviour. She starts zigzagging slowly across the landscape, up and down hedgerows, flying low and frequently landing to investigate holes and grass tussocks. The general appearance is of someone looking for something important, and she is: this is nest-site searching behaviour.

Many species – especially the Early and Garden Bumblebees and the carder bumblebees – tend to nest above ground, and will be looking for thick tussocks of rank grasses or thick blankets of moss over not-too-wet ground. Tree Bumblebees look for tree cavities or their artificial equivalents, bird nestboxes and house eaves. For several species – particularly the Buff-tailed, Red-tailed and White-tailed Bumblebees – an old rodent burrow is ideal, a cool underground chamber with nest material already in place. Bumblebees' ability to see well into the ultraviolet spectrum of light comes in useful here: mouse urine fluoresces in ultraviolet light and bumblebees can follow the trail like an arrow leading back to the nest. Compost heaps and cavities in rockeries or dry stone walls will attract numerous species; the Heath Bumblebee (*B. jonellus*) has even been recorded nesting in an old Long-tailed Tit nest suspended in vegetation.

Desirable nest sites can be a scarce commodity and late-off-the-mark queens will sometimes try to take over the early-stage nest of another queen. A queen has done well to get this far, but to set up a nest and successfully raise the next generation will be a whole new struggle.

Below: The Tree Bumblebee often nests in old bird nests, as here.

At Home with the Bumblebees

When the queen finds a nest site to make her own, her next task is the same whether it's a hole in the ground, an old vole nest, a bird box or just the tangled bottom of a grass tussock. She must begin to build the nest that will shelter her and her offspring for the rest of her life, turning it from a cold, wet hole to a warm, dry home as quickly as possible, fuelled only by flowers.

Setting up the nest

Opposite: A Buff-tailed Bumblebee (*B. terrestris*) queen with her worker bumblebees in the nest.

Different bumblebee species have different ideas of what makes an ideal nest location, but most bumblebee nests start out similarly on the inside. The queen burrows into her chosen material – dry grass and moss, an old bird's nest or rodent bedding are all favourites – and hollows it out, forming a sphere slightly smaller than a tennis ball, with a space in the middle about the size of a ping-pong ball. This chamber will be the focus of the queen's existence for the rest of her life.

From glands in her abdomen, she extrudes wax from between the plates that make up the exoskeleton of her abdomen and, using her mandibles and legs to tease it into shape, she sculpts a shallow waxen cup in the centre of the nest – her first brood cell. Taking regular foraging trips outside to refuel, the queen builds up the sides until the brood cell is about the same size as her body.

At this stage her foraging trips change: she begins to collect pollen as well as nectar, filling the baskets on her hind legs with pollen wetted with small amounts of nectar to make it stick together.

Below: Queens will only fill their pollen baskets once they've established a nest.

Above: Common Carder Bumblebees in their mossy nest.

When she arrives back at the nest, she scrambles onto the brood cell and uses her middle legs to scrape the contents of her pollen baskets into the cell. She then tamps down the pollen with her mandibles, making sure she packs in as much as possible; this pollen will be the only food her first offspring have until they're fully grown adult bumblebees foraging for themselves, and the more nutrition they have as larvae, the bigger and more capable they will be as adults.

The queen will also be busy building herself a thimble-like honeypot from more of her wax flakes. This will usually be near the entrance, and close enough to the brood cell that she can stick her tongue into the honeypot whilst brooding her offspring, sat atop the cell. She fills the pot with nectar brought back in her honey stomach. Filling it can take dozens of foraging trips and hundreds of flower visits, but it will be worth it as this is her insurance policy against bad weather that might prevent her foraging.

Eggs at last

Once the brood cell is full of pollen, the queen will lay her first eggs. Her two ovaries each contain four ovarioles (the small tubes leading from the egg-producing cells to the oviduct, the egg-laying tube) and each ovariole contains either one or two mature eggs at a time. This affects her clutch size: species such as the Common Carder (*B. pascuorum*) that have a single egg per ovariole, can lay eight eggs in a clutch; while species such as the Buff-tailed Bumblebee that have two eggs per ovariole can lay clutches of up to 16 eggs at a time. Cuckoo bumblebees, adapted to smash-and-grab egg-laying raids on nests, can have 6–18 ovarioles per ovary so can lay much bigger clutches at once. Even these clutches pale in comparison to the queen honeybee, who has 160–180 ovarioles in each ovary in order to maintain her vast worker force.

Several eggs are laid onto the top of the pollen in the brood cell. These are pearly white, oval in shape and around 3mm (⅛in) long. Bumblebees, like most other Hymenoptera, have an ovipositor (an egg-laying tube)

Below: White-tailed Bumblebee (*B. lucorum*) nest, showing hatched cells.

Above: The bumblebee's ovipositor (egg-laying tube) is now adapted and used solely as a defensive weapon.

but, as in many bee species, this has adapted into a sting instead, and eggs are laid from the tip of the abdomen, at the base of the sting rather than at its tip.

Once the eggs have been laid, the queen secretes more wax from her abdomen to build a roof across the top of the brood cell and seal in her offspring with their food. Like many other insects (including butterflies and moths) bumblebees have a four-stage lifecycle, with separate egg, larva, pupa and adult stages. The larval stage is key – while adults eat largely to keep themselves going, larvae eat to grow.

This is the toughest time in the queen bumblebee's year, and many nests fail at this point. Her eggs will hatch after four to six days, and the larvae will pupate 10–20 days later, emerging as new adults around a fortnight after that. During this month, the queen must forage as well as keeping the nest temperature high – if the temperature drops to 10°C (50°F) or below, the developing larvae will die, and nests are generally kept at 25–32°C (77–90°F). To achieve this, the queen periodically sits astride her brood cell and buzzes her wing muscles, heating up the inside of her thorax. This warmed blood is then pumped around her body, where it warms the hairless 'brood patch' on the underside of her abdomen; this is pressed against the lid of the brood cell, which quickly gains a semi-melted impression of the queen's underside in the lid.

Boys or girls?

Above: Nests this large are likely to be producing males, queens and workers.

The queen bumblebee and all the workers in the nest are females – males are only produced towards the end of the colony's life. To achieve this, the queen can control the sex of each egg she lays. Unlike in humans, where sex depends on whether an individual has two X (XX) or one X and one Y (XY) sex chromosomes, whether a bumblebee is male or female is determined by the total number of sets of chromosomes an individual has; a system called haplodiploidy. Males have one set of chromosomes, and females (both queens and workers) have two sets. This means that unfertilised eggs will hatch into males, as they contain one set of chromosomes from the mother. Fertilised eggs, with one set of chromosomes from the mother and a second set from the father, will be females.

This results in some very weird (from a human perspective) family trees – males have no fathers and can have no sons, but have grandfathers and grandsons. It makes workers more closely related to their sisters (who share 75 per cent of the same genes) than they would be to their own daughters (who share just 50 per cent of the same genes). It's a major reason why bee nests work – workers pass more of their genes on to the next generation by helping to produce siblings (the new queens) than by laying their own eggs.

The queen mated the previous summer, before hibernation, and stores the sperm in a little sac, the spermatheca, by the ovaries. For each egg that she lays, she chooses whether to fertilise it (for a worker or a new queen) or not (for a male). To fertilise it, she opens the muscle connecting the spermatheca to the egg-laying duct as the egg passes; to leave it unfertilised she leaves it closed. Sometimes unmated females lay eggs (workers when a queen dies, for example), and all these eggs produce males as they can't be fertilised.

Larvae

Bumblebee larvae are white, vaguely C-shaped grubs, and rather resemble maggots in appearance. From the moment they hatch, their sole purpose is to eat, and eat, and eat. They're entirely helpless, with very limited powers of movement, and have no eyes, antennae or legs.

They can only survive by essentially living in their food – pollen. For most animals this would quickly get

Above: Honeybee brood comb, with white grubs curled up inside the cells, waiting for workers to feed them.

very unhygienic, but bumblebee larvae avoid this in a very simple, if unusual, way. They have no anus – instead they have what's known as a blind gut, with an entrance (the mouth) but no exit. This allows them to live in a heap of pollen without contaminating the food that the queen has laboured long and hard to collect.

After two to three weeks of solid munching, and shedding their skins three or four times, the larvae are fully grown. Those destined to be queens (produced later in the year) are larger, even at this stage – they're fed more food more frequently as larvae, and consequently consume around three times as much pollen as males and workers do during their respective developments.

The first batch of workers, however, will be small – with just the queen foraging for them, these are likely to be the smallest bees produced by the colony, a fraction of the size of the queen.

The last thing that they do as larvae is to spin a tough silken cocoon, secreted from glands near the mouth, as a shelter to pupate in. As they pupate, they also finally void the waste from their larval development, keeping it safely hidden away in the base of the cocoon, away from their

Above: The bumblebee brood comb – larvae tucked away.

siblings' food supply. By this stage, the queen has usually laid her next batch of eggs, triggered by the presence of pupae in the brood cells, and there will be a range of larvae, from large to small, grazing alongside each other.

The pupa is white, and in shape closely resembles the adult bee that it will soon become. Over the 14 days between the larva pupating and the new adult emerging, the grub's body will be almost completely broken down into a glutinous soupy mixture, and rebuilt into the familiar bulky shape of the bumblebee it will become. As the emergence date approaches, the features and colouration of the bee inside become visible, looking ghostlike through the thin pupal case.

Newbees!

After five days as an egg, three weeks as an eating machine, and two weeks as soup inside the pupal case, it's finally time for each of the queen's offspring to face the world as an adult for the first time. Once the new adults have forced their way out of the pupal case itself, they bite their way out of the cocoon surrounding it, clamber out, and crawl across the floor of the nest to the honeypot for a drink of nectar.

Below: A bee pupa. The shape of all the adult bee's appendages are already visible as, inside, the larva is broken down into soup and rebuilt as an adult.

At this stage they're fairly soft and floppy, as their exoskeletons take a while to dry off and fully harden from exposure to air. Their wings are soft, crumpled up and pressed against their backs – these too need time to harden up, and will not totally dry for at least a day. Most strikingly of all, the newly hatched bees are covered in almost completely silvery-white hair, which gradually darkens to their adult colours over the next day.

Once their thirst has been quenched from the honeypot, the new workers tuck themselves up by the queen to dry off. They've done well to get this far: of all the eggs laid by the queen, only around 70 per cent hatch; of the larvae, only 75 per cent make it to pupation; and only 90 per cent of the pupae hatch successfully. That means less than half of the eggs laid by the queen make it to adulthood, and their hard work is just beginning.

Worker bees

Above: White-tailed worker bumblebees quickly take over foraging duties from their queen.

A day or so after emerging from their pupae, the new worker bees are dried off and have coloured up. For the first week or so after hatching, the workers can secrete wax from between their abdominal plates, providing the queen with raw materials for building cells. These 'nest bees' help the queen as she devotes her time to producing more workers. Building upwards and outwards from the original brood cell, she constructs new brood cells on the shoulder of the first, forming a brood comb. She also builds more honeypots, and in some cases, a wax cover over the brood comb (the involucre), meanwhile she lives off nectar taken from the communal honeypots.

Once they're ready to go outside for the first time, the new workers take over the foraging duties from the queen, visiting flowers for nectar and pollen to sustain their younger sisters. More or less as soon as they begin bringing pollen loads back, the queen stops foraging herself, instead remaining in the nest full-time. Workers forage and fill the brood cells and honeypots as soon as they're built.

Not all the workers will leave the nest to forage, though most in the first batch will. Some, particularly those that are small or have weak or crumpled wings, will stay in the nest

Above: A queen Common Carder Bumblebee in the nest with a worker.

with the queen and carry out 'household duties'. This isn't a strict definition as these bees don't have the specialised nest-bee roles of honeybee hives – where individuals have regimented roles at different stages of their adult life – but there are a variety of tasks that need to be carried out in the nest to keep everything running smoothly.

These household duties include the removal of dead larvae or adults, defence of the nest against intruders, and maintaining the nest's temperature and humidity balance. Bumblebees try to keep the nest around a constant 25–32°C (77–90°F), and higher temperatures can heat-shock colonies, harming or even killing eggs, larvae and pupae. This can be a particular problem on warm summer days and for surface-nesting species, where the nest is in direct sunlight. On warm days workers bite open temporary vents in the wax cover over the brood comb and tease open holes in the nest material to dissipate heat; workers also act as air-con units, using their wings to gently fan cooler air from the entrance across the brood comb. This behaviour can be particularly evident with Tree Bumblebees (*B. hypnorum*), which often nest in bird boxes. On hot days the entrance holes of their nests can be ringed with workers wafting air inside like miniature fans.

Pollen-storers and pocket-makers

Now fully established, the colony enters its main growth phase. With a squadron of workers foraging non-stop, the volume of pollen and nectar being brought back to the nest is hugely more than the queen managed on her own, meaning each new larva gets more food than any of the first batch did and the workers in each batch are bigger than those in the last (as long there is food to be found). Foraging is a dangerous business, and the mean adult life expectancy of a worker is four to six weeks – this can drop to two to three weeks in the peak season, as the pressure of more hungry mouths takes a toll. Consequently there is a constant need for more workers just for replacement of losses, let alone colony growth.

Above: A more-advanced pollen-storing species, the Red-tailed Bumblebee (*B. lapidarius*).

Different bumblebee species carry out their larvae-rearing in slightly different ways. Pocket-making species such as the Common Carder Bumblebee and the Garden Bumblebee (*B. hortorum*) build wax 'pockets' at the base of the nest. Workers fill these with pollen and larvae eat their way through it in much the same way as the first batch of workers did. This means that larvae are in competition for food and may do better or worse depending on their position within the pocket (larvae on

Below: A pocket-making Common Carder Bumblebee.

the edge have less pollen available and generally do worse). This means there's often considerable variation in size of workers in these species, even within a batch.

More advanced species are known as pollen-storers. In these species (such as the Early and Red-tailed Bumblebees, *B. pratorum* and *B. lapidarius*), the queen lays eggs into empty cells and the developing larvae are fed individually by workers as they return to the nest. These species also have specific pots to store excess pollen in (hence 'pollen-storers') before it is fed to larvae. This individual feeding means larvae are not in direct competition with each other for food, so usually get fairer shares, consequently workers tend to be less variable in size.

Above: A worker Garden Bumblebee. She will empty her pollen baskets into a large communal pollen store.

Below: Pollen-storing bumblebee species tend to all have similar-sized workers, as here.

Rebellion

Above: Honeybees often kill their queen when they rebel against her.

As the workers get bigger, they become physically capable of laying eggs of their own. Because of bumblebees' haplodiploid sex determination system (see page 47), these unfertilised eggs hatch into males, which consume pollen as they develop but which don't forage as adults, meaning a net loss to the colony.

The queen, aiming to maximise her own reproductive output, acts aggressively towards fertile workers, trying to bully them into non-egg-laying submission. In honeybees, the queen produces a pheromone that suppresses worker ovarian development so for honeybees aggression towards fertile workers is a last resort. In bumblebees, which are less advanced socially, bullying is the first resort and there is a clear link between ovarian development and increased levels of aggression.

Eventually, as worker numbers build, the nest reaches a stage where the queen is incapable of bullying all the workers all of the time, and some of the larger workers develop ovaries and begin laying eggs. In response to this, the queen begins to spend time exploring the higgledy-piggledy brood comb, looking for eggs which she didn't lay and eating them.

Some workers may even 'rebel' against their queen, beginning to eat the queen's eggs while they lay their own – the Red-tailed Bumblebee is particularly prone to this behaviour. An increase in the number of workers secreting wax to build the nest causes subtle changes in the chemical composition of the wax. Workers use these changes to gauge the colony's life-stage and when the threshold is crossed they become agitated, racing around the nest, attacking each other, and destroying existing eggs and replacing them with their own.

Fortunately for the queen, this tends to be fairly late in a colony's life, and can be viewed as a fallback position in case of the old queen's death – if she dies the nest will still produce males, and so still has a chance of producing a new generation.

Below: The emergence of the first male Red-tailed Bumblebees often sparks rebellion back in the nest.

Drones and princesses

Above: The North American species *B. bimaculatus* - one of the few queens to be helped by her sons as well as her daughters.

The ultimate purpose of a bumblebee nest is to produce more queen bumblebees, who carry the genes of the queen and her colony to a new generation. The many hundreds of workers that the queen may raise in a lifetime are simply a means to that end: a way of stockpiling sufficient pollen that the new reproductive generation will get off to a good start in life.

After a few batches of workers have been produced, and the nest's stocks of pollen and nectar are increasing nicely, the queen lays her first unfertilised eggs, which will turn into males. These are reared in the normal way, but they generally eat less pollen as larvae than do workers. This allows the colony's food stocks to increase further, just in time to allow bigger, more pollen-hungry new queens to be reared. Handily, this also means that males (drones) leave the nest some time before their new queen sisters, and as workers do not allow them back into the nest once they've left, they disperse further afield, which reduces the chances of mating with their sisters and the resultant inbreeding that would occur.

In some North American species – *B. griseocollis, B. pennsylvanicus, B. affinis* and *B. bimaculatus*, at least

Below: A male Tree Bumblebee. Barred from re-entering the nest, much of his time is spent resting on flowers.

Above: A queen Great Yellow Bumblebee (*B. distinguendus*) looking for a suitable nest sites.

– males will also spend some time incubating the brood cells before leaving, a rare example of paternal care in insects. It is possible that this behaviour may be more widespread in bumblebees, but it's very difficult to study in the wild.

What makes a queen a queen, rather than a worker, is not entirely understood, and seems to vary slightly between species. Essentially, though, fertilised female larvae that are fed more often, or for longer, turn into queens, and equivalent larvae fed less turn into workers. In some species, such as the Buff-tailed, White-tailed (*B. lucorum*) and Red-tailed Bumblebees the queen emits pheromones for most of her life that inhibit workers from feeding larvae as frequently, or as much per feed, as they potentially could. Larvae fed under this regime turn into workers, therefore the queen can build up large worker forces. Buff-tails have the biggest nests of any British bumblebee with about 400 workers.

When she decides to produce new queens, the old queen switches off her pheromone production, which allows workers to maximise the feeding rate. The larvae that develop with as much pollen as they can take reach adulthood as queens. Queens generally take around three times as much pollen to develop to adulthood as an average worker and many nests – up to 75 per cent in bad years – do not manage to successfully rear new queens.

Queens that do reach adulthood spend most of their time in the nest at first. They feed on the remaining stored pollen and nectar to build up their fat reserves for the coming winter, only foraging outside the nest if they have to. By this time the old queen is usually dead or dying (the reproductive lifespan of a queen, from laying her

first eggs to death, is around 18 weeks on average) and, with no matriarch and no more larvae to feed, the nest organisation begins to crumble.

When the new queens are full of pollen, they fill their honey stomachs with nectar from one of the communal pots and leave the nest in search of a male to mate with and a nice safe hole in which to ride out the winter. They will never return to their natal nest again.

Above: A mating pair of Red-tailed Bumblebees.

Left: A queen Tree Bumblebee hibernating in a tree crevice.

Flowers, Foraging and Feeding

Bumblebees are made entirely of the stuff of flowers, and when she's outside the nest, the average worker bumblebee is on a never-ending foraging mission. Flitting from flower to flower, she must fill her pollen baskets and honey stomach and return to the nest again and again, feeding the immobile larvae and then heading back to the flowers.

Why do bumblebees forage?

The vast majority of the time, foraging bumblebees are collecting pollen to feed to the larvae, and nectar for energy for themselves. The high protein content of pollen (variable between plant species but usually around 25 per cent, comparable to beef, lamb or peanuts) makes it ideal for feeding to fast-growing larvae. The adults have a limited need for protein but, instead, have an almost unlimited appetite for sugar to keep their huge wing muscles supplied with energy. In bee-pollinated plants, nectar is often more than 30 per cent sugars. By contrast, nectar from plants pollinated by hummingbirds is usually about 10 per cent sugar, neatly demonstrating how energy-intensive bumblebee flight really is.

Neither pollen nor nectar evolved specifically to encourage insect pollination, let alone as a reward for bees. Pollen is the male sex cell for flowering plants and also for conifers. Conifers, which have been around for more than 300 million years, are wind-pollinated; they release huge clouds of pollen into the environment in the hope that at least one pollen grain will end up in the right place – watch a large conifer on a windy day in autumn to see just how much a single tree can produce! The origin of nectar is less clear, but it is also thought to have pre-dated the evolution of flowering plants.

Opposite: Long or complex flowers, like this deadnettle, are especially suited to bumblebees.

Below: The high protein content of pollen has also led to its use for human consumption in products such as pollen lollipops.

Collecting pollen

Above: Bumblebees can get smothered in pollen inside flowers.

Flowers have become irrevocably linked with pollen, nectar and pollination, but in fact even flowering plants have diverged down two different main pathways when it comes to pollination. Wind-pollinated species like grasses reproduce in the same way as conifers, producing lots of pollen to drift on the wind and hopefully land on their small, green, female flowers. The pollen-producing male flowers are still visited by bumblebees – it's pollen, after all – but there's no incentive for the bees to visit the female flowers of wind-pollinated species.

Below: Hairy faces are perfect for pollen transfer.

The majority of flowering plants developed flowers to minimise the amount of pollen they had to create, by using animals to ferry pollen between flowers on their behalf. Bumblebees are adapted for the efficient collection of pollen in several ways. Their general hairiness is not just good for surviving the cold, but also acts like Velcro: pollen grains that slide off smooth, hairless wasps and beetles get caught in a bumblebee's fur. Bumblebees also generate an electrostatic charge as they fly, which helps pollen stick to them in a similar way to how a balloon sticks to a wall after it has been rubbed on a jumper.

Most bee species carry pollen in a brush of hairs on the legs or abdomen, but bumblebees (and honeybees) are different. A female bumblebee's modified hind tibia (the shin) has been adapted to form a pollen basket. The tibia is shiny, flattened and broadened out so much that it's almost concave, and it's fringed with long, curved hairs that form the sides of the basket when it's full of pollen.

Above: Solitary bees carry loose pollen in a brush, seen here beneath the abdomen of this Red Mason Bee (*Osmia rufa*).

At the bottom of the pollen basket, the joint between the tibia and the topmost section of the tarsus (the bumblebee's 'foot') is adapted to collect the pollen from between the bee's hairs. As the bumblebee grooms itself with its hind legs, the pollen is packed together at the base of the pollen basket. As the bee visits flower after flower, the stored pollen builds up until it completely fills the pollen basket. Bumblebees generally return to their nests carrying around 25 per cent of their body weight in pollen, but they are capable of carrying 75 per cent of their body weight (or more) when they need to.

Below: Bumblebees carry damp pollen in pollen baskets.

Collecting nectar

Above: Borage – one of the best plants for bumblebees.

It's not just pollen that's carried back to the nest: nectar is ferried back as well, to top up the nectar pots that provide food for non-foraging adults such as the queen. Even larvae are fed some nectar, mixed in with the pollen. While everyone who's watched a bee visit flowers will have seen pollen collection in action – grains heaping up on the bee's hind legs – nectar collection is less obvious.

Like honeybees, bumblebees have a honey stomach: a bag at the front of their digestive system that fills the front part of the bee's abdomen. This is gradually filled via a succession of sips of nectar from flowers. As she forages, the bee consumes enough to keep going by allowing a trickle of nectar to flow from the honey stomach into her digestive system. She usually drinks around 10 per cent of what she collects, and brings home about 0.3ml (½ teaspoon) of nectar per foraging trip.

As the nectar sits in the honey stomach, it's gradually dehydrated to save space, as water is removed through the walls of the stomach. This is the same process by which honeybees make honey, but bumblebees don't take things that far – bumblebees' stored nectar is usually at least 20 per cent water, compared to around 17 per cent in 'true' honeybee honey.

Below: Small, clustered flowers are great for bees.

How to forage

Above: Flowers that have abundant nectar can attract several bees at once.

The queen will forage for nectar for herself alone until she establishes her nest; after that, she and her daughters collect both pollen and nectar according to need. In summer, workers will be out foraging from dawn to dusk. Light is a limiting factor: bumblebees navigate using visual landmarks and can't find their way home in the dark. It's not uncommon to find workers that stayed out too late sleeping in flowers.

To leave the nest, forage and return successfully, bumblebees need to be pretty good navigators. When they leave the nest for the first time, they create a mental map of the surrounding landmarks during an orientation flight, spiralling around and outwards from the nest. On subsequent flights, workers memorise the position of the sun relative to the nest entrance. When it's time to come home, their accurate internal body clock allows them to work out how much time has passed, and so what angle

Above: Bumblebees are usually the first pollinators to be up and about each day.

they need to fly relative to the sun to get back to the nest. Bumblebees' eyes can detect the plane of polarised light, so they can tell where the sun is even when it's hidden by clouds. They can then guide themselves over the last bit of their journey by navigating using the landmarks they memorised the first time.

Bumblebees have been recorded foraging up to 10km (6¼ miles) from the nest, but it's unlikely that this is normal – the forager would need to spend a lot of time feeding herself just to get back to the nest, and wouldn't be able to make many round trips. Many species are known as 'doorstep foragers': individuals usually forage no more than about 400m (1,300ft) from the nest, and make many return trips in a day.

Foraging activity usually peaks around midmorning, and on hot summer days there is a definite midday lull, as workers return to fan the nest or just rest to prevent overheating. Actively flying individuals must have

Above: As you might expect from their name, Corpse Lily flowers smells of rotting flesh, which helps to attract flies.

their flight muscles above 30°C (86°F), and 35–40°C (95–104°F) is more usual. Unfortunately, proteins such as muscles begin to degrade not far above 40°C (104°F) and a bumblebee flying on a hot day is likely to begin to cook itself from the inside out.

Bumblebees have a very good sense of smell and can even detect the scents of flowers like Viper's-bugloss which are scentless to humans. It's our great good fortune that people and bees are attracted to the same floral scents. If bees, like many flies, were attracted to the smell of rotting meat it's unlikely that gardening would be as popular as it is!

When a bee finds a flower by following the scent, it homes in on 'nectar guides' on the petals. These are ultraviolet markings grown by the plant to guide foragers to the right part of the flower. To us they're invisible, but to bumblebees, able to see much further into the ultraviolet spectrum, they stand out clear as day.

Bumblebee communication

Unlike honeybees, bumblebees don't dance to tell their nestmates where food can be found, but there is communication between workers. Some of this is passive: a bee sampling high-quality nectar brought back to the nest is likely to go out foraging herself, often even following the bee that brought back the good nectar. This is especially likely if nectar stocks are low.

Some bumblebee communication is more active. Workers returning to the nest with high-quality nectar release a pheromone that stirs the other workers into action and more go off to forage. This pheromone may also help bees to learn floral scents, particularly those released from the nectar as the successful worker deposits it into the communal nectar pots.

Bumblebees also communicate via 'excited runs'. Successful worker bumblebees returning to the nest run and scuttle at speed around the nest and over the brood comb, buzzing their wings; the more successful they've been, the quicker they run and the longer they run for, before flying away. These runs again stir any nestmates into foraging action, especially when nest supplies are low.

Below: Honeybees 'waggledance' but bumblebees use less sophisticated communications.

Which flowers to visit?

Above: Red Clover is a bumblebee superfood.

The most important thing for a bumblebee colony is to have enough flowers to visit throughout the flight season, roughly February/March to September/October in the UK. This is why farmed fields, even of flowering crops such as Oilseed Rape or beans, are not much use to bumblebees. Their flowers are useful while they're there, but these crops are only in flower for a few weeks and, to a bumblebee, are huge empty dead zones the rest of the year.

Outright flower diversity at any one time is not necessarily what the bees are looking for either. Various studies have found that bumblebees have strong preferences, and spend the vast majority of their time visiting a few flower species. In the UK, 92 per cent of bumblebee flower visits in flower-rich field margins were to six plant species, and 80 per cent of the pollen collected by 15 bumblebee species across a wide range of habitats came from just 11 plant species. However, there's also evidence that larvae fed on a diet of pollen from several different plant species do better than those fed mostly on pollen from any one plant species.

Below: Wind-pollinated cereal crops are of virtually no use to bumblebees.

What do bumblebees need from pollen?

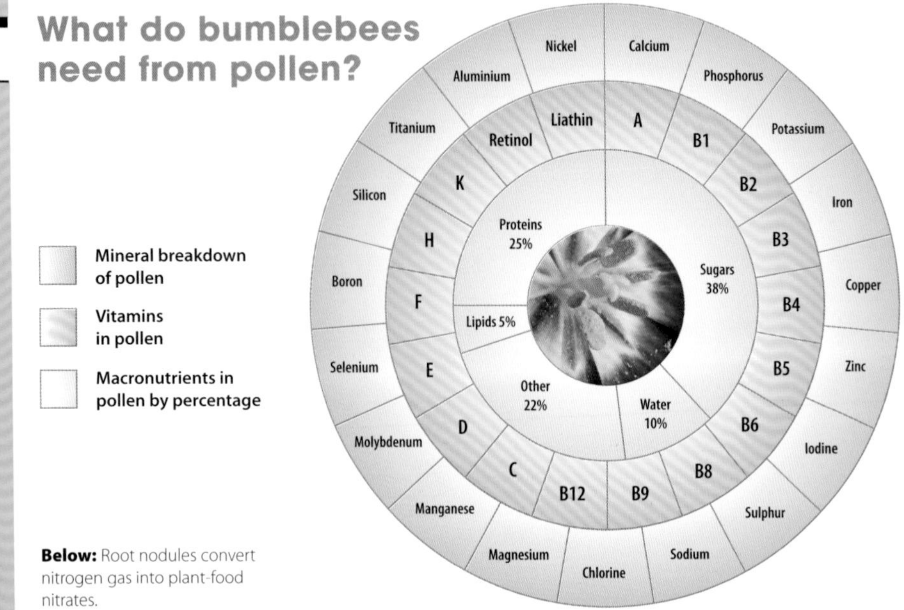

Below: Root nodules convert nitrogen gas into plant-food nitrates.

Bumblebees are also known to medicate themselves, killing their internal parasites by visiting plants (for example, the tobacco plant) with toxic or harmful compounds in the nectar that they usually avoid.

Red Clover is a real bumblebee superfood, and up to 30 per cent of a colony's flower visits can be to this species; up to 40 per cent can be to clovers in general. The family they're part of – Fabaceae, the peas, beans and legumes – seems to be particularly important to bumblebees. A study of bumblebees in Britain and New Zealand found that, of the top 20 most visited flower species, eight were in the family Fabaceae. This is no accident. Many plants in this family host bacteria in nodules on their roots, which can convert nitrogen from the air into ammonia or nitrates. These help the plant grow, but also mean that they produce nectar that is unusually rich in amino acids, making it a useful secondary source of protein for bumblebees and their larvae.

Flower preferences

Different species of bumblebee will preferentially visit different plant species, usually because of particular physical limitations. Bumblebees are generally divided into long-tongued and short-tongued groups. Long-tongued species, such as the Garden or Common Carder Bumblebees (*B. hortorum* and *B. pascuorum*), are able to feed on flowers where the nectar is a long way from the opening, like complex-shaped dead-nettles or long, tubular Foxgloves. Long-tongued species preferentially visit these long-corolla flowers as they have less competition at those plants: most other flower-visiting insects, including the short-tongued bumblebees, can't reach these flowers' rewards, and instead visit more open flowers like Sunflowers.

Other flower preferences are size-related. The Buff-tailed Bumblebee (*B. terrestris*), the UK's largest species, is big, heavy and not very agile. Consequently, it tends to

Above: Long-tongued species, such as the Garden Bumblebee forage from long flowers.

Above: Open flowers are available to all pollinators.

Left: The acrobatic Early Bumblebee (*B. pratorum*) hangs upside-down from flowers.

Above: Long-tongued bumblebees are perfect for pollinating Foxgloves.

avoid pendulous flowers or small, delicate blooms that can't take its weight. Instead it likes to visit horizontal or upwards-facing flowers, especially chunky ones that provide a substantial landing platform. Another large species, the Red-tailed Bumblebee (*B. lapidarius*), also likes horizontal flowers, but prefers species like daisies, where lots of flowers make up a single flower head, as the bee can walk between them.

The Early Bumblebee is a spring specialist, and has to be able to make use of all the small number of flowers that are out at that time of year. This very agile species is even able to visit flowers that face vertically downwards. Early Bumblebees can often be seen dangling off comfrey flowers. Although they have relatively short tongues, they have narrow heads so can use fairly tubular flowers by sticking their whole head into them.

Individual preferences

Bumblebees can also detect when a flower is worth stopping on. If a flower has recently been visited by another bee, it's likely to be empty. Like most other animals, bumblebees leave a chemical footprint where they stand, and this lingers after they've gone. Bees visiting the same flowers at a later time are able to use this smelly footprint to tell how long ago the visit was. Flowers also have a pattern of negative charge on their petals, which is disrupted for several minutes after a bee lands on them. These systems, and the bees' knowledge of the flower species' refill time, helps them guesstimate whether a flower is worth probing for nectar.

There are also individual-level differences in the flowers used. While some flowers, like flat, disc-shaped dandelions, are simple and easy for any insect to use, others are more complex and can be difficult for insects to visit successfully. Bumblebees have to learn how to get at the pollen and nectar hidden within these complex flowers. Having invested the time learning how to use the flowers, some individual workers

Below: Bumblebees face more competition on flat, open flowers.

Above: Big pollinators rub the right bits of the hollyhock flower to collect pollen.

Left: Bumblebees can assess the current state of the flower from chemical footprints left on the petals by other bees.

then show a preference for visiting the same type on each foraging trip, even bypassing other flower species that could contain more food.

As a learned preference, this is different from inbuilt preferences for characteristics like flower colour – individual workers from the same nest may learn how to use different flowers, and so show markedly different preferences. These preferences also suit the plants. For pollination to be successful, a flower's female parts (the stigma) must receive pollen from the same species of flower. This is far more likely to be achieved when the pollen-carrying insect specialises in a single flower species.

Nectar robbing

One effect of the evolution of long, complex flowers has been to lock out short-tongued bumblebees. These can't physically reach the nectar by reaching though the flower, but they don't let that stop them. Bumblebees have large, strong mandibles and are easily capable of biting through the back of flowers, directly above the nectary. Once the hole is made, they can insert their tongues and drink as much nectar as the plant has to give. In a field of beans, a single foraging bumblebee can bite 2,000 holes in a day. Some species are well adapted to the 'nectar-robber' role: the European bumblebee *B. wurflenii* has mandibles with large 'teeth', all the better to bite flowers with.

The benefit may seem one-sided as there's no chance of pollination, but this is not always the case. Much of the time, plants set as many seeds, of the same quality, whether or not nectar robbers were active during flowering. This is presumably because enough visitors used the flowers in the 'correct' way, even with others robbing the flowers – even nectar-robbing bumblebees collect pollen, after all.

Some studies even show positive effects on the plant, probably because nectar robbing decreases the amount of nectar available to non-robbing foragers. This forces them to visit more flowers on more plants, which increases the chances of pollination by pollen from a different plant, lessening any inbreeding and generally letting the plant produce more or better seeds.

Below: Short-tongued bumblebees often resort to using their large mandibles to bite into flowers, opening access holes closer to the nectar source.

Pollination

From a bee's perspective, spreading pollen between flowers – pollination – is a missed opportunity: every grain of pollen that fertilises a flower is one the bee could have taken back to the nest. Pollination is not a happy coexistence; it's an arms race, with plants and bees grudgingly needing each other but trying to maximise their benefits for the minimum of effort.

Plants want to move just enough pollen from the anthers (male parts) of one flower to the stigma of another for pollination to occur. Energy spent producing pollen that goes elsewhere, or producing nectar, flowers and so on, is wasted unless it contributes to this goal. Bumblebees want to be completely efficient about their food collection. Ideally they'd fill their pollen baskets and honey stomach with as few flower visits as possible, without losing any en route. The complexity of pollination – nectar production, flowers of assault-course complexity, pollen baskets and the rest – are all ways that the two sides have tried to beat the system over millennia, to move and countermove as each tries to minimise costs and maximise rewards.

The Foxglove is a classic example of the arms race in action. The plant produces a spike of flowers, and every day a new flower opens at the top of the spike, and one withers and dies at the bottom. New flowers, at the top of the stem, are male, and produce pollen. As they age, and move towards the bottom of the flower spike, they become female, receptive to pollen, and produce more nectar and less pollen. Bumblebees start at the bottom of the spike and move up, drinking nectar and getting progressively more covered in pollen as they reach the male flowers. Realising it's not getting much nectar any more, the bee flies off to the Foxglove plant next door. Again it lands at the base of the flower spike, this time smearing pollen grains over the fertile stigmas as it crawls down the long tubular flower to reach the nectar. The bee has a full honey stomach and pollen baskets, the flower has been pollinated, and both sides win.

Above: Bumblebees and solitary bees (the latter seen here) are the supreme pollinators.

Threats

Bumblebees and their ancestors have survived the worst that the past 100 million years could throw at them. They rode out the impact that drove the dinosaurs to extinction, and used the continental collision that created the world's highest mountain range as a springboard to success. As with any successful group, a whole ecosystem of predators, parasites and diseases has grown up around the bumblebees, but it's only in the past century that things seem to have gone seriously wrong.

In Britain, three species that once bred here have gone extinct. A further six species – a full quarter of the species we have left – have declined to the point where they are officially recognised as species of conservation concern. Across much of the UK, especially the 'central impoverished zone' of the English Midlands, only the 'Big Seven' species can be reliably found.

It's not just Britain where bumblebees are struggling. The official global IUCN Red List has two North American species (*Bombus franklini* and *B. affinis*) listed as Critically Endangered, one step away from extinction. The Patagonian Bumblebee (*B. dahlbomii*) has just become the first South American bumblebee to make the Red List. In Europe, Cullum's Bumblebee (*B. cullumanus*) has followed extinction in Britain (see page 16; last seen 1941) with a reduction in range across the whole continent, and is now listed as Critically Endangered.

Above: The Shrill Carder Bumblebee (*B. sylvarum*), Britain's rarest species.

Left: The declining Moss Carder Bumblebee (*B. muscorum*).

Opposite: While a few species such as this Red-tailed Bumblebee (*B. lapidarius*) remain common, many others have declined considerably.

Nest predators

Above: Watchful Common Carder (*B. pascuorum*) around the nest.

For some predators – those that can avoid or ignore the stings – a bumblebee nest is a juicy, high-value meal full of pollen, nectar and immobile larvae. For other predators that can sneak in undetected, it's an all-you-can-eat buffet with free refills. Either way, a nest defended by up to 400 workers armed with venomous stings and able to detect intruders by vision or smell is not an easy target. Some bumblebees, such as the North American Yellow Bumblebee (*B. fervidus*), will defend their nests not just with stings and jaws but also by daubing sticky nectar over the attacker, gumming it up.

Natural predator

One of the main threats to bumblebee nests comes from close to home. The UK is home to six cuckoo bumblebee species that, unlike true bumblebees, don't build nests and produce workers. These are a specialised group that have lost the ability to collect pollen or feed their own brood, and consequently they have no worker caste. Instead, they're adapted to a piratical home-invader lifestyle of taking over the nests other species have built. Each 'cuckoo' has a preferred host species or two, but are capable of attacking other, usually closely related species.

Above: Red-tailed Cuckoo Bumblebee (*B. rupestris*) is parasitic on the Red-tailed Bumblebee (*B. lapidarius*).

Cuckoo queens start the year with a lie-in, emerging six weeks or so after their preferred host species. After feeding themselves up for a couple of weeks, they begin the quest for a nest. Spending their time lazing around on flowers, they watch for queens of their host species and trail them back to their newly established nests. They can also sniff out nests themselves over short distances, ideally looking for a nest where the first batch of workers has just hatched. That way there's a labour force available for the cuckoo, but not too many defenders.

Once she has found a nest, the cuckoo will try to sneak inside. She usually spends a few days at least under and around the brood comb, even burrowing down into the nest material to disguise herself with the scent of the host nest. If approached by workers, she may play dead, but in many cases this is not enough and workers often sting her to death, or at least evict her. Large, strong nests are frequently littered with the corpses of cuckoos.

If she survives this, the cuckoo then seeks out the host queen. Armed with stronger mandibles and a longer sting than the 'true' bumblebees, and with thicker skin acting as armour, cuckoos are formidable opponents. Usually the host queen is killed, stung in the neck or through the joints between her abdominal plates. Sometimes she survives, relegated to the role of worker by the now-dominant cuckoo, who eats any eggs she may try to lay.

The cuckoo, now queen of all she surveys, also subdues any resisting workers with a combination of pheromones and threatening behaviour. She doesn't kill them if at all possible, as they will be the labour force out foraging for the nest. To ensure that all this effort goes to rearing her own larvae, the cuckoo kills all the eggs and young larvae she can find in the brood cells. She spares the older larvae that are about to pupate and become workers in their turn.

Above: A melanic male Field Cuckoo Bumblebee (*B. campestris*).

Having lost the ability to produce wax of her own, a cuckoo queen must build brood cells by recycling the wax already in the nest. She can lay 20–30 eggs at a time, far more than the host 'true' bumblebees, so her batch of offspring is soon complete. She may have to defend her eggs from workers rebelling against her, and she will also patrol the nest and eat any eggs that may be laid by these rebellious workers. 'True' bumblebee eggs are smooth, whereas the eggs laid by cuckoos are heavily ridged, so the cuckoo tests any eggs she finds with her tongue, and eats any with smooth surfaces.

With no foraging to be done, the queen cuckoo's destructive work is finished. She will abandon the nest, leaving the workers to continue to forage to provision her offspring. In time they, both males and females, will emerge and leave, and can frequently be seen lazing on umbellifers and other flowers, drinking nectar without the frantic flower-visitation of the nest-building species.

Infiltrators

Left: The Large Velvet Ant (*Mutilla europaea*).

It's not just cuckoo bumblebees that try to take advantage of the bumblebee's hard work building up the nest. Large Velvet Ants (*Mutilla europaea*, actually a wasp species) have a thick, armoured exoskeleton and a powerful sting (in the USA the family are known as 'cow-killers'). They sneak into the nest, lay their eggs on the cluster of brood cells and the larvae feed on bee pupae.

Nests are also targets for the fly *Brachicoma devia*. This grey-black, hairy fly is another species that attacks the bumblebee brood, invading nests in spring and laying live larvae into the brood cell. These larvae attack the immobile final-stage bee larvae, draining them of fluids like miniature Draculas, before dropping down to the bottom of the nest and pupating amongst the debris. A week later, the new adults emerge and the cycle repeats.

Even moths get in on the act. Bee Moths (*Aphomia sociella*), Wax Moths (*Galleria mellonella*) and Lesser Wax Moths (*Achroia grisella*) all feed on the brood comb within bee and wasp nests. The Bee Moth can be a particular problem for bumblebees: larvae begin by eating old wax and other detritus, but as they grow they increasingly feed on bee pupae and larvae. Living in amazingly tough tunnels made from silk that they spin themselves, they're protected against the stings of patrolling workers.

Above: A male Bee Moth (*Aphomia sociella*).

Nest raiders

Right: Wood Mouse; a threat to overwintering queens and small nests.

It's not only the sneaks that nesting bumblebees have to look out for. Plenty of mammals will destroy nests, eating the brood and stores. Rats, shrews, mice and other small rodents are deterred by stings and present more of a threat early on, when there are few, if any, workers. They also eat a lot of overwintering queens that didn't manage to hide themselves away well enough.

Badgers are champion nest eaters and can sniff out a well-established nest before using their powerful front legs to excavate and tear apart the nest material. Thick skin around their noses and eyes protects them from most stings, and they can reduce an active bumblebee nest to a crater full of dazed bees in no time at all.

Below: Badgers will dig out and eat large nests.

Mid-air dangers

Outside the nest, there are plenty more threats to foraging bumblebees. Most spider webs are too puny to contain a bumblebee in a hurry, but a few species, such as the Wasp Spider (*Argiope bruennichi*), spin strong webs that can ensnare an unwary worker bee.

A few impressively agile insect predators can even catch a flying bumblebee in midair without a web. Hornets and large species of dragonfly catch smaller workers on the wing. After grabbing them, the predators use their lethal jaws to snip off the wings and legs of their victims before eating their meaty body.

Above: The Wasp Spider's strong web can easily catch bumblebees.

Below: Bumblebees make an ideal snack for a foraging Hobby.

Birds also happily eat bumblebees. Hobbies, flycatchers, shrikes and bee-eaters are all known to catch them in flight, though most prefer honeybees or solitary bees.

A few species have a worse fate in store for any bumblebees they manage to catch. Both parasitic wasps in the genus *Syntretus* and flies in the family Conopidae wait on flowers for a bumblebee to approach. When one does, they fly up to it and grab the bee in mid flight, swiftly injecting an egg between the plates of the bumblebee's exoskeleton, then letting go and flying off before the

Above: Conopid flies wait on flowers for a bumblebee to parasitise.

bee knows what's happening. These eggs hatch into larvae inside the bee where they absorb nutrition from the bumblebee's blood before eating its internal organs. Eventually the larvae kill the host and pupate, ready to repeat the cycle again. Conopid flies even seem to be able to mind-control their host bumblebee as, when the fly larvae inside it are mature, the bee digs a shallow hole – in the ground immediately before dying. The larvae then pupate inside the dead bee at the bottom of this hole more sheltered than if the bumblebee had simply expired above ground.

These parasites don't get it all their own way, however. In the case of the conopid fly *Sicus ferrugineus*, bumblebees have been seen apparently fighting back. The parasitic larva needs higher temperatures to survive and grow than the bee does to survive. Parasitised bumblebees have been recorded not returning to the nest at night, instead staying outside, where it's colder. This slows the development of the parasitic larva, sometimes so much that the larva dies and the bee can return to work.

Flowery threats

Above: Crab spiders, *Misumena vatia*, have potent venom.

On flowers another eight-legged peril awaits. The crab spider *Misumena vatia* can change colour to match the flower it's on, and lurks beneath the flower head. When a potential meal alights, the spider leaps out fangs first, ready to use potent venom that can kill even an insect as large as a bumblebee in seconds.

Sometimes the flowers themselves can be dangerous. The flowers of some lime trees contain narcotic nectar, and bumblebees that drink too much of it become slow and drowsy, or even die in extreme cases. Blue Tits and Great Tits are clever little birds and have learned both the association between lime trees and dozy bees, and the best way to kill a sleepy bumblebee without being stung. They peck open the thorax or abdomen, leaving the exoskeleton but eating the organs or muscles. Consequently the ground beneath lime trees can be covered with hollowed-out bumblebees, especially in spring.

Left: Blue Tits are adept at pecking bumblebees from flowers.

Bumblebee mites

Above: Mites are harmless hitchhikers on this queen bumblebee.

Bumblebees – especially queens – often have little orange-pink mites on them, particularly around their wing bases. There are plenty of tutorials online about removing them but there's really no need. These mites are in the genus *Parasitellus* and, despite the name, they're harmless to the bee and are beneficial to the nest.

They don't suck the bee's blood – the mites on the bees are at non-feeding larval stage. Instead they eat detritus in the nest – old wax, bits of dropped pollen, faeces, mould – and keep the place much cleaner than it would be without them.

They only use the bees for transportation, grabbing hold of a new queen as she leaves the nest in summer, holding tight all winter, and then dropping off her once she has a nest site of her own. Bees have the ability to carry heavy loads of pollen and nectar, so a few extra micrograms makes little or no difference – a small price to pay when the end benefit is an army of nest cleaners that stop the whole place going mouldy.

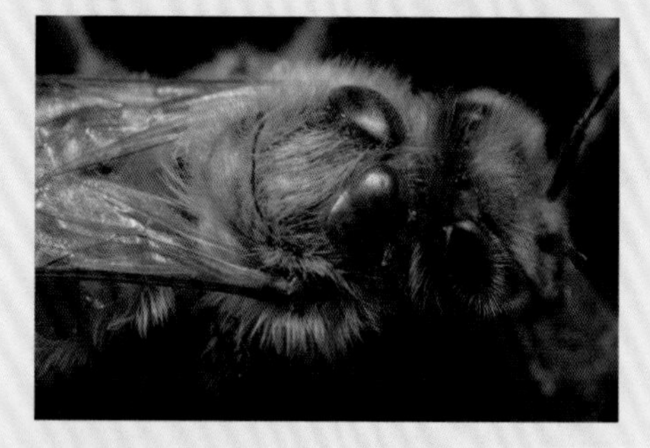

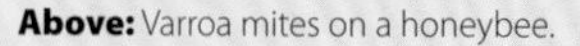

Above: Varroa mites on a honeybee.

The Varroa Mite (*Varroa destructor*), the mite linked to honeybee die-offs, does not multiply on bumblebee broods and needs winter warmth to survive – the annual lifecycle of bumblebees does not suit it at all.

Internal threats

A wide range of organisms can call the insides of bumblebees home, and parasites and diseases are becoming increasingly recognised as a problem for bumblebees. A nematode worm – *Sphaerularia bombi* – infects queen bumblebees from the soil during their winter hibernation. The parasite takes over the bee, shutting down her nest-building instinct and laying up to 10,000 eggs inside the host. These hatch, and are shed with the faeces before the queen returns to her overwintering site and dies, infecting the area with more eggs and larvae.

Above: A honeybee infected with Deformed Wing Virus.

Other microscopic infections, including *Crithidia bombi*, *Apicystis bombi* and *Nosema bombi*, are (as the names suggest) specialised on bumblebees. They all tend to make bees less efficient at feeding themselves, foraging for the nest, direction finding and flying generally. Although they may not kill an individual bee outright, infections have a colony-level effect, eroding the efficient functioning of the nest, reducing the amount of pollen and nectar stored, and ultimately reducing the number of queens that the nest is able to produce.

Some diseases have only recently been found in bumblebees after decades as honeybee diseases. Deformed Wing Virus, Kashmir Virus, Acute Bee Paralysis Virus and *Nosema ceranae* have all recently been found in bumblebees. The long-term effects of these on bumblebees are not yet known, including whether they can transfer between bumblebees or only from honeybees to bumblebees when the two meet on flowers. However, with their smaller numbers of workers per nest, bumblebees are potentially at risk even if the diseased individuals just die and do not also spread the infection.

Striped canaries in the coal mine

Above: In butterflies, such as these Common Blues (*Polyommatus icarus*), each adult male and female is an active member of the breeding population. This is not the case with bumblebees.

Bumblebees' unusual colonial lifestyle makes them more resilient to threats in some ways, but much weaker in others. In most animal species all adults can breed, therefore the adult population equals the breeding population. In bumblebees, though, most adults are sterile workers, and each nest has just one reproductive individual, the queen, for much of the year, so the breeding population of bumblebees is much smaller than the adult population as a whole. This is good in some ways – loss of workers is not an immediate drop in the size of the breeding population, for example – but it has one major drawback. For a breeding population that is large enough to be sustainable in the long term, with a minimum of about 30–50 queens, bumblebees must also produce workers. Around one hundred times as many workers in fact – about 5,000 individuals.

This means that colonies need a lot of food, and they need it all the way through the flight season, from March to September in the UK. The more flowers there are in the area, the more efficient bumblebees' foraging will be, with shorter return trips between the nest and the resources. In general, each colony needs an area of about 10 sq km (4 sq miles) to forage over, so a viable population size of 30–50 nests needs a large area of flowers. By contrast, a breeding population of 30-50 butterflies is just 15-25 adult female butterflies and 15-25 adult males, which is a much lower food requirement for the same size of breeding population.

This all means that healthy populations of bumblebees need large, well-connected areas of good habitat to thrive. As habitat areas shrink or become fragmented the bumblebees vanish, the choosiest species disappearing first.

Habitat loss and degradation

The importance of this increased food demand becomes clear when you look at what's happened to the wider countryside over the past century. Between 1932 and 1984, more than 90 per cent of the UK's natural grassland – a key bumblebee habitat – was destroyed. Between 1945 and the present, that figure is thought to have reached 97 per cent. Clover-rich horse pastures have disappeared and hay meadows have become silage pastures, cut and fertilised so often that flowers cannot set seed. Over the same post-war period, 160,000km (100,000 miles) of hedgerows have also been lost. With them have gone nesting sites, overwintering sites and important early food sources in Blackthorn and willow trees.

It's the same story elsewhere. In the US, the state of Iowa was once 85 per cent prairie grassland, ideal bumblebee habitat. Now less than 0.1 per cent of it remains.

Below: Flower-rich meadows have become rare sights in the UK.

It's not just about the outright loss: much of what is left has been gradually degraded and, from the bees' perspective, is of lower quality than it was a century ago. Monocultures create feast-or-famine situations where acre after acre of crop produces flowers for just a couple of weeks but is useless for the other 50. Ever-increasing fertiliser use favours grasses over flowers, while herbicide finishes off many of the rest.

Above: Shrill Carder Bumblebee, now restricted to a handful of sites across southern England and Wales.

In 1928 an entomologist wrote in his notebook '*Bombus sylvarum* everywhere as usual'. Nowadays every sighting of the Shrill Carder is a red-letter day. Bumblebees need flowers and flowers need bumblebees, yet we are systematically stripping out flowers from the wider countryside. We are creating an extinction vortex, with mutually dependent plants and bees spiralling after each other to their deaths.

Below: Hedgerows just aren't enough habitat to sustain bumblebees.

Neonicotinoids

Above: This bumblebee may suffer after-effects from this meal.

Using insecticides – neonicotinoids or anything else – kills insects. That much is agreed, but it's difficult to go much further into the pesticide debate without controversy.

Neonicotinoid insecticides are mostly used as a seed dressing on crops such as Oilseed Rape. As 'systemic' insecticides, they are spread throughout the plant and incorporated into it. This makes every part of the plant slightly toxic, including the pollen and nectar, but only with a very low dosage of the insecticide. Neonicotinoids can kill bumblebees outright, but only at doses much higher than those encountered in flowers in the field.

What is becoming increasingly clear, however, is that exposure to these low doses, particularly chronic exposure, has noticeable effects on bumblebee health. These sublethal effects particularly affect brain function, reducing bumblebees' abilities to navigate, forage successfully and return to the nest. This in turn reduces the nest's ability to produce males and new queens at the end of the season.

Nature's Little Pollinators

Bumblebees are under threat. They're declining, and if they go, it's amazing what they'll take with them. A huge majority of crop and wild flower species depend on insect pollination: without bumblebees and other pollinators, food and the countryside as we know it will simply cease to exist.

Pollination services

Nature can be seen as providing 'ecosystem services' to people. These can be large-scale and irreplaceable (for example, plants providing oxygen) or small-scale and localised (for example, enjoyment of a particular view), or anything in between. Pollination is a classic ecosystem service, provided by anything that moves pollen from one flower to another, and it is increasingly used as an additional justification for conserving bees and other flower-visiting insects.

For good reason, too. A huge amount of scientific research in recent years has enabled us to work out fairly accurately the financial value of pollination to humankind. Globally, animal pollination (not just by bees) is responsible for around 35 per cent of global crop production, and around 80 per cent of wild flowers and 84 per cent of high-value crops are dependent on insects and other animals for pollination.

In the UK, the value added to crops by insect pollinators has risen steadily from £440 million/year in 2010 to £691 million/year in 2015; even back in 2008, insect pollination added £3.4 billion to the European economy, and this is likely to have increased considerably since. Although it is possible for humans to pollinate crops it would involve an army of people walking the fields armed with paintbrushes, dabbing pollen from one flower to the next. This was conservatively estimated to cost £1,500 million/year in 2010.

Opposite: Worker bumblebees spend their entire adult lives visiting flowers.

Below: Crops like Sunflowers need insect pollinators.

Above: Honeybee hives are often positioned between crops that need pollination.

As this is almost four times the value of the crop, the price of insect-pollinated food – apples, for example – would have to quadruple to remain profitable.

Loss of pollinators will not lead to a global food shortage as many staples, such as wheat and barley, are wind-pollinated. It would, however, remove the good stuff from our diets – the apples, tomatoes, oranges, pumpkins, Bakewell tarts – and leave us with bland staples of bread, couscous and other cereals.

Bumblebees are particularly important as they will forage in colder conditions than most other pollinators, including earlier and later in the year or the day. Additionally, 8 per cent of the world's flowers can only be fertilised by sonication (also known as 'buzz pollination'). These flowers – including tomatoes and blueberries – produce pollen inside tubular, pepperpot-like anthers. For pollen to be released, a bee must vibrate the anther hard enough to dislodge the pollen, and it will then sprinkle out. In Britain, only bumblebees can do this, by vibrating their wing muscles at a high frequency as they clasp the anther.

Above: Orchards don't produce anything without insect pollinators.

There's often talk of honeybees, and of how the gradually increasing number of hobbyist beekeepers means that we don't need to worry about wild pollinators. If only it were that easy! Honeybees account for only around 7 per cent of the flower visits across rural and urban areas, including farmed land and nature reserves. At most, honeybees can provide only a third of our crop pollination needs and there are hundreds of species – like the buzz-pollination flowers or night-flowering species – that they don't pollinate at all. To survive, plants need a diversity of flower visits, from insects with different flight periods, visitation times, sizes, shapes and tongue lengths. We cannot maintain the diversity of plants in our countryside if we don't maintain the diversity of pollinators out there too.

China's Apple Valley

Above: Hand-pollination is time-consuming and fiddly.

Maoxian County in south-west China was the apple-growing capital of the country. Apples need lots of pollinating insects, especially bees, to transfer pollen between trees in order to grow high-quality fruit. But in Maoxian County the orchards were sprayed eight times a year, every year, for 40 years with a cocktail of pesticides, herbicides and fungicides. Along with the loss of habitat from squeezing in ever more apple trees, wild pollinators were wiped out in the region.

With no bees around, from the early 1980s the farmers of Maoxian County had to take over and do the job themselves. With just a five-day window to pollinate the trees, armies of farmers, their families and seasonal workers would collect pollen from each tree into bags, then dab it with a paintbrush into the flowers on other trees. By the year 2001, every tree in Maoxian County was hand-pollinated.

Each person could only do 5–10 trees a day, so this got very expensive as wages increased. Honeybee hives were tried, and although they were eight times cheaper, the hives kept dying out from the witches' brew of sprayed chemicals.

Increasing wages and the decreasing price of apples soon made hand-pollination unsustainable. Nowadays Apple Valley is down to 30 per cent apple crops and falling – an entire industry destroyed by the loss of wild bees.

Repair and recreate

Above: Grassland flowers like knapweeds are great for bumblebees.

By far the biggest thing bumblebees need is a reversal of the habitat loss that has occurred over the past century. The good news is that this is beginning to happen. Farmland takes up around 70 per cent of the UK's land surface, so improving life for bumblebees here is vital.

Agri-environment schemes pay government and European money to farmers to conserve and improve the conservation value of their land. One of the main activity options that farmers can choose to take up has for many years been to plant 'pollen and nectar strips' in their field margins. By 2009, more than 4,000ha (9,900 acres) of bumblebee-friendly margins were planted, providing a late-season boost for existing bumblebee nests to last longer and produce the next generation of males and queens.

These have been specifically aimed at bumblebees and the bees have responded well. In particular the Ruderal Bumblebee (*Bombus ruderatus*) showed a similar decline to the Short-haired Bumblebee (*B. subterraneus*) during the 1970s and 1980s but, while the Short-haired went extinct in 1988, the Ruderal has made a good comeback, feeding in the pollen and nectar margins.

Other agri-environment scheme options available to farmers include restoring hedgerows. This increases early-season bumblebee food from Blackthorn, Cherry Plum and willow, and provides more nesting and overwintering sites for queens. Even minor tweaks, such as changing the month that meadows are mown or planting flowering shrubs, can make a real difference to bumblebees – that few extra per cent they need to produce queens at the end of the season instead of just males, or to tide the queen over an Easter cold snap, stopping her freezing to death in her snowed-in nest. Organisations such as the

Bumblebee Conservation Trust are working with landowners to put the flowers back into the countryside, and the bees – gradually – seem to be starting to follow. But it's a long, hard road and there are plenty of potential roadblocks ahead.

One of the most important bumblebee habitats is, surprisingly, old brownfield sites. These rubble-strewn, man-made sites, particularly around the Thames Estuary, tend to be covered with the kind of plant that rare long-tongued bumblebees crave – vetches and trefoils. Sites such as the Crossness Pumping Station, managed by Thames Water, can hold acre after acre of the bright yellow flowers of Narrow-leaved Bird's-foot Trefoil in summer. Dotted in amongst them are the flitting figures of hundreds, if not thousands, of bumblebees, including the rare Brown-banded Carder (*B. humilis*) and Britain's rarest bumblebee, the Shrill Carder (*B. sylvarum*). The Shrill Carder has a characteristic high-pitched buzz and on a warm summer's day you can lie back in the flowers with your eyes closed and still know when this special little bumblebee is nearby.

Above: Thistles can be covered with bumblebees in midsummer.

Below: Complex or long flowers are almost exclusively accessible to bumblebees.

Garden habitat

Above: Pollinator-friendly planting can be spectacular.

The other important concerns for bumblebees are that they have enough food from March to October and that there are connections between areas that are suitable for them. It's not just farmland and nature reserves that are important: gardens are a vital habitat for many bumblebee species (and are visited by many other species) and verges can provide essential bee-friendly flight lines through inhospitable areas. The seven most widespread bumblebee species can be more abundant inside gardens than outside them.

Anyone with access to a garden can join in with habitat creation. The Bumblebee Conservation Trust has a website – Bee Kind (beekind.bumblebeeconservation.org) – where you can provide a list of the flowers in your garden and the BCT will tell you where the gaps are for bumblebees (maybe there's not much nectar provision in June, for example). They will also suggest a range of plants to fill those gaps and give your garden a score for its bumblebee-friendliness – why not compete with your friends? The Royal Horticultural Society also has a list of more broadly pollinator-friendly plants – Perfect for Pollinators – which is usually shown on the plant labels themselves.

The key points are to avoid overbred flowers – double flowers and that kind of thing. Even if the bumblebee could force its way in through the squished-together mass of petals, many produce no pollen or nectar – to a bumblebee they are distracting false advertising. Most simple cottage garden varieties are good, as are tubular flowers (they tend to be more specialised for bumblebee pollinators). Wildflower mini-meadows can be great as well – at least leaving the clover and dandelions in the lawn long enough to flower or only mowing one half of the garden at a time. Bumblebees' needs are relatively simple!

Above left and above: Both roses, but only the left flower is useful to bumblebees.

Left: Bumblebees will happily use non-native flowers.

Bumblebee introductions

Above: Fast travel (comparatively!) to the Antipodes allowed the shipping of bumblebees.

The knowledge that bumblebees are fantastic pollinators is not new. During the 19th century European farmers in New Zealand were struggling – they needed Red Clover growing in the fields for their livestock, but the long, narrow flowers of clover couldn't be pollinated by any New Zealand species, not even by the European honeybees the farmers had also brought with them. Consequently the plants never reproduced, and the farmers had to spend time and money buying in seed from Europe and replanting the clover time after time.

The advent of the new steamships in the 1880s – fast, and carrying refrigerators – opened up another possibility for New Zealand's beleaguered farmers: import some bumblebees to pollinate the pastures. In 1885, 93 over-wintering bumblebee queens were shipped from the UK to New Zealand, followed by a further 143 in 1906. Four species survived: the Buff-tailed (*B. terrestris*) and Ruderal Bumblebees, now ubiquitous across the country, and the Garden (*B. hortorum*) and Short-haired Bumblebees, which are present but with a more restricted distribution.

Below: New Zealand's productive agriculture – made possible by bumblebees.

The bumblebees did the trick and within five years New Zealand was producing so much Red Clover that it had become a net exporter of seed. The meat and dairy industry grew and grew on the back of the bumblebees' foraging. The UK, the market for 90 per cent of New Zealand's production, gained easy access to cheap meat and dairy that was vital to its survival, particularly during World War Two.

There are downsides to shifting species around the world, however. Following their agricultural success in New Zealand, Ruderal Bumblebee queens were introduced to Chile in 1982–3, again for Red Clover pollination. They were later joined by Buff-tailed Bumblebees, also deliberately imported as pollinators, and both species have since spread out across South America – the Buff-tailed spreading almost 200km (124 miles) per year. Unfortunately, they weren't alone when they were imported. At least one of the species – almost certainly the Buff-tailed – was carrying the tiny parasite *Apicystis bombi*. While this doesn't seem to affect the European species, unfortunately the native Patagonian Bumblebee (*B. dahlbomii*) has no defences against it. When the host is infected, the parasite weakens it, causing increased worker deaths and decreased colony formation.

Above: Stands of bumblebee-pollinated red clover.

The toxic combination of competition and disease from the European bumblebees has seen the Patagonian Bumblebee retreat from much of its former range across South America to now be largely restricted to Tierra del Fuego. It is currently the most threatened bumblebee on Earth.

Left: The 'flying mouse' – a queen Patagonian Bumblebee.

Commercial bumblebees

In light of the problems caused by importing bumblebees into South America, it's concerning to see the spread of commercial bumblebees. Sellers rear bumblebee nests in captivity and sell them to anyone who needs a quick burst of pollination. Globally, more than two million bumblebee colonies are reared every year, and the industry was worth in excess of £44 million in 2006.

Most commercially reared bumblebees are Buff-tailed Bumblebees, although other species are also bred less successfully in captivity. They're reared in huge sheds, fed pollen collected from honeybees, and manipulated (by changing the temperature, light and carbon dioxide levels) to produce nests all year round. Used in greenhouses and polytunnels to pollinate strawberries, tomatoes, sweet peppers and the rest, they've become a vital cog in the machine that makes out-of-season produce possible. Less aggressive than honeybees as well as being individually better pollinators, they're also far cheaper than the £8,000 per ha (£3,200 per acre) per year that it cost even back in 1988 to manually pollinate each tomato flower using a vibrating rod (the 'Electric Bee').

Below: Buff-tailed Bumblebee, the main commercially-reared species.

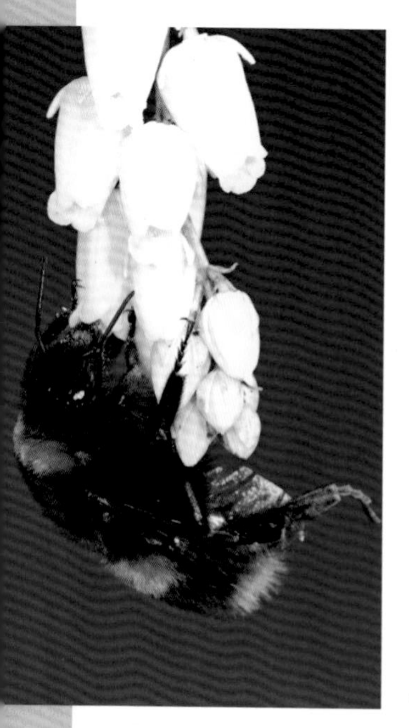

And if they stayed there – safely tucked away in the greenhouses and polytunnels – all would be well. Unfortunately, they don't. Studies have found that up to 73 per cent of the pollen in commercial bumblebee nestboxes in greenhouses has actually come from plants only growing outside the greenhouse. Increasingly, commercial bumblebee nests are also used in open fields, or even in gardens, and this is a problem. As happened with the Buff-tailed Bumblebees introduced into South America, many commercial bumblebees and their nests have been found to be contaminated with diseases and parasites, both in the bees themselves and in the pollen they're fed on and have stored.

In Britain, these diseases have been found to be more common the closer you get to users of commercial bumblebees. Abroad, bumblebees native to North America were reared in Dutch commercial facilities before being

Above: A nest box of commercially reared bumblebees being used for crop pollination.

returned home, where commercial rearing continued in facilities in the USA. Very soon, bumblebees in the California factory began suffering badly from heavy infections of the parasite *Nosema bombi*, which increases mortality and decreases egg production and colony size. At the same time, wild populations of several common and widespread native bumblebee species began to decline, including the Western (*B. occidentalis*), Yellow-banded (*B. terricola*) and Rusty-patched (*B. affinis*) Bumblebees. The few molecular analyses of these samples suggested the infection was a recent introduction.

Additionally, introduced bumblebees tend to visit introduced plant species. In New Zealand, the four introduced bumblebee species have been recorded visiting 400 non-native plant species and only 19 natives. In Tasmania, 83.5 per cent of the flower visits by the introduced Buff-tailed Bumblebee were to introduced weed species, inadvertently helping the weeds to spread at the expense of native species.

Watching Bumblebees

Often the best way to watch bumblebees is to attract them into your garden – with the added benefit that you'll be helping save the sound of summer! Saving the bees is something that anyone can help with. Urban areas, with their mix of gardens, parks, tiny nature reserves and neglected but flower-rich 'lost' areas, can be very good for bumblebees. There are 15–20 million gardens in Britain, covering about 4,000 sq km (1,500 sq miles) – an area the size of Somerset.

In particular, most people's desire to have their gardens alive with flowers for as long as possible matches the bumblebees' need for flowers from at least March to October, though it's important that the flowers do actually provide pollen and/or nectar. The pretty but useless overbred flower varieties, with nectar locked away behind an impenetrable shield of petals, or with an inviting appearance but no supply of pollen or nectar, are worse than useless for bees, merely acting as time-consuming distractions.

Opposite: Varied and flower-rich, gardens cover more of Britain than nature reserves.

Below: Formal gardens can be just as good for bumblebees as gardens planted with wildlife in mind.

Subletting the garden: nesting bumblebees

While providing food is vital to bring bumblebees to your garden, it's important to remember that, for the future of the species, nest sites are also essential. Bumblebees aren't aggressive, even when defending the nest, and can be watched from a metre (3¼ft) or so away without harm to you or the bees. By the time a nest is noticeable they tend to be at their peak, with a steady stream of workers going in and out, and there's usually only about a month of activity left – so enjoy it while it lasts! However, one species to be slightly wary of is the Tree Bumblebee (*Bombus hypnorum*) – this species seems to be slightly more aggressive in nest defence than many other species and, in particular, seems not to like buzzing vibrations such as drilling or lawnmowers, too close to the nest.

In general, though, a bumblebee nest is a great thing to have in the garden. Increasingly, people have been trying to provide nesting habitat by putting artificial nestboxes – 'bumblebee houses' – in their gardens. Unfortunately, unlike solitary bee homes, bumblebee houses are far from a sure thing. Most tests in Britain have found only 0-2.5 per cent of the artificial bee houses were occupied.

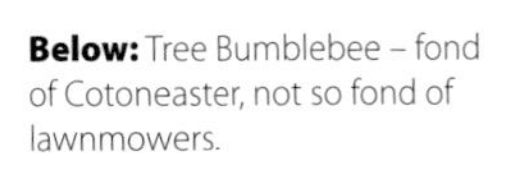

Below: Tree Bumblebee – fond of Cotoneaster, not so fond of lawnmowers.

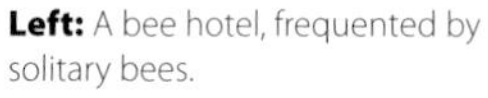

Left: A bee hotel, frequented by solitary bees.

Luckily, there's no need to buy a bee house to provide nesting habitats in your garden. If you have cavities in a rockery or similar, or even in an upturned flowerpot under a garden shed, it's worth just tucking a handful of dry moss or kapok inside as a kind of 'nest starter kit'. If you can use some bedding from a pet rodent, so much the better. Several species of bumblebee like to nest in old mouse or vole nests in the wild – in fact, it's always worth looking for bumblebee nests near Hazel trees, as the nuts attract rodents, who then make nests that the bumblebees reuse.

Below: Hazelnuts, a favourite of small rodents.

Making a mini meadow

Above: Yellow Rattle keeps grasses down, and attracts bumblebees in its own right.

Probably the best thing to do, if you have space, is to leave an area of at least 2m by 5m (6½ by 16½ft) as a mini meadow of rough grassland. Only mow roughly once a year in September/October, and rake off the cuttings – otherwise they'll accumulate into a thick thatch that will stop anything germinating. This mini meadow, left to its own devices, will provide nest sites for surface-nesting species like the Common Carder Bumblebee (*B. pascuorum*). When flowers start to come through (Dandelions, clover, trefoils and similar: naturally or planted), it will provide foraging habitat for any bumblebee species (and much more besides).

You can even use plants that are semi-parasitic on grasses, such as Red Bartsia or Yellow Rattle. Not only are they great bumblebee food, but they'll also slow down the grass growth, providing an opportunity for other flowers to come up without being smothered by long grass.

A rough-grass mini meadow will also be good for voles, which will dig the holes and make the nests that then provide nest sites for underground-nesting species like the Buff-tailed Bumblebee (*B. terrestris*). Studies in parks in San Francisco have found that the abundance of bumblebees matches the abundance of rodent holes in the parks – more hole-digging voles means more hole-nesting bumblebees.

If you can lay a piece (50cm x 50cm/20in x 20in or bigger) of roofing felt or corrugated iron on part of the mini-meadow, even better – bumblebees will nest underneath it, and beetles, toads and all sorts of other wildlife will use it as shelter. After a couple of years it will kill off the vegetation beneath the sheet – this is the ideal time to move the sheet to somewhere new and plant a couple of bumblebee-friendly plug plants in the bare area.

Above: Red Bartsia: another grass-parasitising bumblebee favourite.

Left: Voles inadvertently create bumblebee nest burrows.

Boosting the bumblebee box

If you do go with the artificial bumblebee home, there are a few things you can do to maximise the chances of bumblebees taking up residence. Make sure the nestbox is in a south-facing spot, but is somewhere shady and sheltered – not too exposed to direct sunlight, or the bees will overheat when summer comes. Somewhere at the base of a hedge or beneath a bush is the best bet. Don't plonk it right on the ground – put down a couple of bricks or stones and place the nestbox on top. This will leave a bit of an air gap underneath the nestbox, keeping it warmer on cold days and preventing any water flowing in.

Bumblebees are not the most nimble of flyers, so they like to have a 'runway' to land on at the front of the nest, and a tunnel to walk through to actually get into the nest itself. If you prop dry stems around the nestbox, you can use a length of hosepipe or similar

Below: Like cargo planes, heavily laden bumblebees like to have a runway to land in front of their nests.

Above: If you have the right flowers in a garden, bumblebees will turn up.

as a tunnel from the ground up to the entrance – just make sure that no water will flow down the pipe. Using a white stone or similar to mark the pipe entrance is a good plan – the bees will use it as a landmark (and it'll be a reminder not to stand on that bit of the garden).

Once your nestbox is in place, line the bottom with corrugated cardboard (this will help absorb any moisture from the colony), and put a good handful – a mouse nest's worth – of dry, fibrous material in a clump at the bottom. If it's slightly rodenty, so much the better. Bumblebees can't bring in their own nest material, so this is a key step. Put the lid back on, retreat to a safe distance – and keep an eye out for prospecting queens.

One thing not to do is to buy a box of live bumblebees for your nestbox. Yes, it will guarantee the box is occupied, but as we saw on page 102, releasing these commercial bumblebees risks spreading diseases to the wild bumblebees in your garden. Even the official UK risk assessment says that commercial bumblebees should not be released, only used in sealed greenhouses. Don't do it.

Damsels in distress

Above: Bees often carry a scattering of harmless mites, as on this White-tailed Bumblebee (*B. lucorum*).

If you look at queen bees in spring, especially if you find one looking a bit under the weather and can have a good look at it, you'll usually see little orange-pink mites, generally clustered around the wing bases. As I explained on page 86, these mites are harmless, but all over the internet you'll see advice to remove them – different websites advocate dunking the entire bee in warm water, or flicking the mites off with cocktail sticks, or crushing them with forceps.

Don't do this. Bumblebees are more than capable of flying with a little extra load, and the mites are very helpful back in the nest, keeping things spick and span. The best thing you can do for a grounded queen is to either put her somewhere sunny on a flower, or to give her a teaspoon of sugarwater to drink (before you pop her back outside on a sunny flower). Don't use honey – it's increasingly clear that diseases can be spread from honeybees to bumblebees, including through pollen and other honeybee-associated products, such as honey. Just keep adding sugar to a small quantity of warm water until it stops dissolving. If possible present it to the bumblebee soaked into a small piece of sponge or J-cloth – she'll be able to drink from the saturated material without the risk of getting covered in sticky sugarwater.

Right: A spoonful of sugarwater can help tired bumblebees.

Photographing bumblebees

Above: This isn't a bumblebee high-five – it's telling you to back off.

If you want to get close to bumblebees – to photograph them, for example – the best way is to approach them on flowers, particularly on cooler days when they may be slightly slower than they are when it's hot. Big flat, open flowers, such as umbellifers or Sunflowers, are best – these flower heads are actually made up of loads of tiny flowers packed closely together, so the bumblebees will land and walk across the flower head, drinking as they go.

Don't let your shadow fall on the bee (that's a sure-fire way of scaring off any insect), and don't breathe on it either – in both cases, it will think you're a predator and scarper, sharpish. Move slowly, and keep an eye on the bee's signals. If it lifts a leg off the flower and waves it at you, like a tiny high-five, it's actually a warning – the bumblebee is trying to say 'That's quite close enough, thanks'. If you keep closing in the bee is likely to fly away; if you stay where you are, it's likely to go back to what it was doing.

Below: Often the best way to find bumblebees is to wait at suitable flowers, such as Dandelions.

Above: Carefully used, a net can help you get close enough to the bee to identify it.

There are other ways to get close-up views of bumblebees. Close-focus binoculars are a good way of getting views of bumblebees without having to get too close – some will focus on objects as close as 1m (3¼ft) from your face, so even the bumblebees you're standing next to are fair game. Plus, as with any binoculars, you can turn them round, look down the 'wrong' end, and use them as a field microscope.

Of course, doing this to a bumblebee as it trundles across a flower is difficult. To get a proper close-up look at live bumblebees, you need to catch them. The easiest way is to use a butterfly net – you can sweep sideways across the bee like a tennis stroke. Alternatively, when the bee is on a flower, hold the net like a frying pan, pull the net bag up like a teepee, and gently lower over the bee. When bumblebees take off, they do so vertically upwards and will head straight into the end of the net, where you can grab them.

Transfer the bee into a clear lidded pot and you'll be able to use a hand lens, or your reversed binoculars, to check the smaller details that are often needed to identify bumblebees species. Keep potted bees out of direct sunlight (the pot will act as a greenhouse), and release after a maximum of 15 minutes, to avoid dehydrating them.

ID and bee mimics

There are some easy ones, but often identifying bumblebees can be a tricky business, with many species sharing the same basic colour pattern. For instance, more than half of the British bumblebee species are striped yellow and black with white tails.

The reason for this is mimicry. Bumblebees are brightly coloured in the first place to warn potential predators that they're dangerous, so it makes sense if they all look similar – it minimises the number of patterns that predators have to learn before leaving all bumblebees alone.

This opens a possibility for other species. If they can copy the bumblebee's appearance, then predators will think that they too are best avoided. This strategy has been adopted by a host of other species, especially hoverflies such as *Volucella bombylans*, which even has different colour forms to mimic different species of bumblebee (see page 7). It's even arguable that male bumblebees, which have no sting, are mimicking their more dangerous sisters.

Fortunately, it doesn't take much practice to split the bumblebees from the wanna-bees, and with just seven widespread and abundant bumblebee species you'll get the hang of them in no time. When you're confident that you can identify the bees you see, do record your sightings – these kind of biological records are a window on the world of wildlife, and the best way we have of taking the pulse of the planet. Details on how to record are in the 'Further Reading' section (see page 125).

Above: A hoverfly head – huge eyes and short, stumpy antennae.

Below left: Even when the overall mimicry is impressive, the hoverfly's head doesn't look right.

Below: A bumblebee head – comparatively small eyes and long tubular antennae.

BREWERS SINCE 1828
J.W. LEES
THE
BEE HOTEL

Bumblebees and People

Alongside butterflies and ladybirds, bumblebees form a select band of insects that are liked by almost everyone. From folklore through to the modern day, bumblebees (and honeybees) are well-known symbols of friendliness, industry and thriftiness.

Language

Bees feature more than perhaps any other insects in language and sayings, though most are about bees in the general sense and many of the remainder are specifically about honeybees. Common sayings that are at least in part attributable to bumblebees include 'a bee in your bonnet' and 'make a bee-line for it', and most obviously in the saying 'as busy as a bee'. Other expressions like using 'the bee's knees' to describe something good come slightly more out of left-field. In fact, the saying 'the bee's knee' was first recorded in the 18th century to mean something small and insignificant. The current meaning seems to have emerged from a 1920s craze for animal-related praise for good things, which also included 'the cat's whiskers', 'the monkey's eyebrows' and 'the kipper's knickers'. Not all such expressions have stood the test of time!

Busy
Busy
Busy

Above: The saying 'busy as a bee' was inspired by the ceaseless foraging of bumblebees and honeybees.

Places

A variety of places have been named after bumblebees. In Old English, *'dora'* meant bumblebee, and it is likely that place names such as Dorney, in Buckinghamshire, are from this source.

In the USA, names can be more blatant. In Utah, Bumblebee Spring rises on Bumblebee Mountain, near Bumblebee Lake and Bumblebee Canyon. In California, the hamlet of Bumblebee is on Bumblebee Road,

Opposite: The Bee Hotel, Abergele. As ever, the bee shown is a bumblebee.

near Bumblebee Creek. Neither area has an obvious association with bumblebees (other than a supposed prevalence of bees and wasps in the area) so the origin of the names must remain shrouded in mystery.

The ghost town of Bumble Bee, near another Bumble Bee Creek in the Bradshaw Mountains of Yavapai County, Arizona, does have a foundation myth to explain the names. According to legend, both the creek and the town gained their names in 1870 when a visitor from Nevada, J. X. Theut, was being harassed near the creek by the town drunk, K. Billingsley Callaway. Thinking quickly, Theut threw a rock at a nearby bees' nest and the insects drove the drunk away.

Below: A carved wooden bumblebee at the RSPB's Minsmere reserve.

What's in a name?

In Middle English the word 'bumble' meant to hum or drone, so 'bumblebee' seems an obvious name. Yet bumblebees have not always been known by their current name. The humble bumblebee was once the humblebee, so named because, as it flies, its flight muscles make a humming noise. In 1912, Frederick Sladen wrote *The Humblebee, Its Life-history and How to Domesticate It* but even by then, Beatrix Potter had written about the troublemaker Babbitty Bumble in *The Tale of Mrs Tittlemouse* (1910) making a mossy nest in the burrow of the eponymous heroine. Over the first half of the 20th century 'bumblebee' gradually displaced 'humblebee' as the name of choice. And by the time the first edition of the New Naturalist book on bumblebees was published in 1959, 'bumblebee' had displaced 'humblebee' for good.

A still older name for the bumblebee is 'dumbledor', a name shared with several large black dung beetles and which generally meant a humming or buzzing

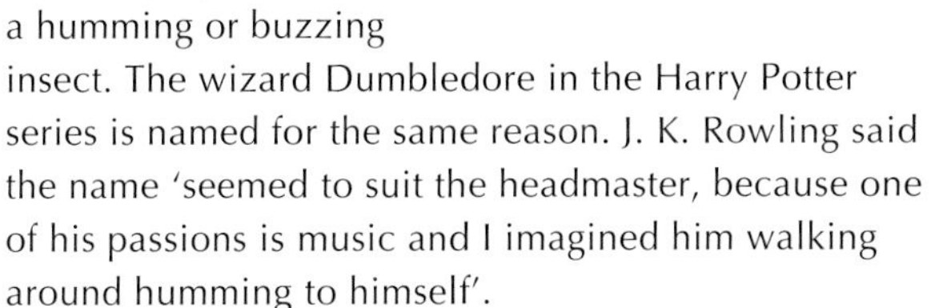

insect. The wizard Dumbledore in the Harry Potter series is named for the same reason. J. K. Rowling said the name 'seemed to suit the headmaster, because one of his passions is music and I imagined him walking around humming to himself'.

Below: Beatrix Potter's story featured the troublesome Babbity Bumblebee.

Myths and legends

From mankind's earliest cave paintings, bees have featured in religion and mythology. Indeterminate species of bee are depicted in rock art from the Palaeolithic era in Spain and the Mesolithic in India. There are depictions of bee gods and goddesses in the Central American Maya civilization and in pre-Hellenic Greek culture, including gold plaques found in Rhodes and dating to perhaps 9,000 years ago. Bees of all kinds have often been credited with the ability to see into the future. This association probably comes from observation of the worker's ceaseless pollen-collecting over the summer, storing food like the ant in Aesop's fable of the ant and the grasshopper.

Below: A cave painting showing a honey gatherer getting attacked by a swarm of bees, Cueva de la Arana, Bicorp.

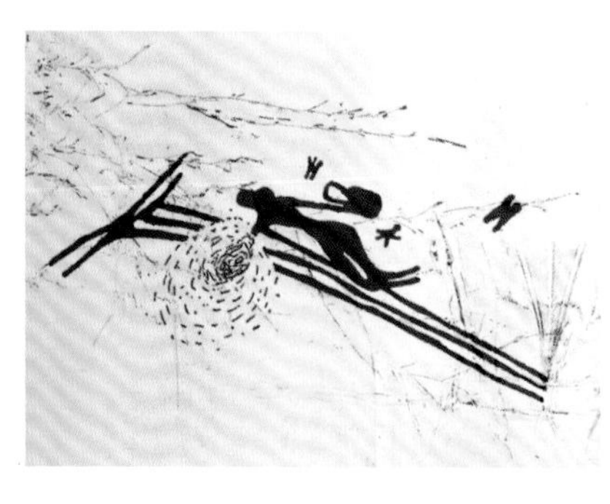

As honeybees became domesticated animals and beekeeping became more important as a trade, depictions and deifications increasingly showed honeybees and beekeepers rather than wild bumblebees. In Egyptian mythology, for example, the tears of the sun god Ra landed on the sand and were transformed into honeybees, while the bowstring of the Hindu love god Kamadeva is made of honeybees.

Folklore

Gradually bumblebees were relegated from symbols of the gods to folklore and old wives' tales, often with an echo of the earlier legends. In Britain, a bumblebee buzzing around your house was thought to foretell the arrival of a visitor. If the bumblebee was a red-tailed species the visitor would be male; if white-tailed, they would be female. History is silent on what an all-ginger bumblebee such as some of the carder bumblebees foretold! Sometimes this was extended: if the bumblebee entered the house and landed on a chair, the visitor would stay only for a short while but, if it landed on the bed, they were likely to stay overnight. What was definite was that if anyone killed the bumblebee, the arriving visitor would bring only bad news.

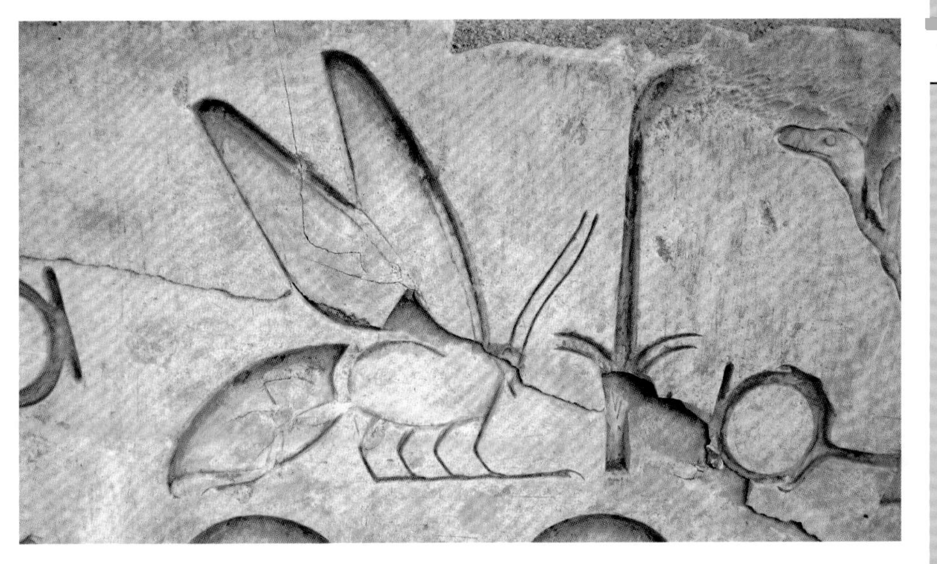

Above: Hieroglyph of a bee from ancient Egypt, where honeybees signified tears of the sun-god Ra.

Bumblebees also had a close association with luck. In Wales, having bees of any kind set up home in or near your house was considered lucky, as they would bless you with prosperity, and to find a bumblebee on a ship was also considered good luck. Having a bumblebee land on your hand meant money would soon be coming to you. A folk charm found in Dawlish, Devon, featured three dead bumblebees in a bag (see photo) and is thought to have been a prosperity charm; the Museum of Witchcraft at Boscastle in Cornwall sells a replica, thankfully with the dead bees replaced by ceramic versions.

Bumblebees – especially black ones – also had an association with witchcraft. At least one Lincolnshire witch supposedly had a bumblebee as her familiar, and black bumblebees were thought to bring bad news, or be a more general bad sign if they entered the house. Killing a bumblebee was thought to always bring bad luck, and having a bumblebee die in your house would bring bad luck and poverty to the household. Swearing at a bumblebee was said to (understandably) drive it away.

Below: A prosperity charm from Dawlish, Devon, now in the Museum of Witchcraft.

Symbol and inspiration

Above: Bumblebee from *Transformers*.

More recently, the bumblebee has become an icon of doggedness in the face of adversity. This seems to result from the notion that scientists once proved that bumblebees couldn't fly but, not being bound by the opinions of others, the bumblebees kept flying anyway. As we saw in Meet the Bumblebees (see page 13), this results from a misconception (it was actually proved that bumblebees would struggle to *glide*, but that is of little importance since they flap their wings), but it continues to provide a handy title for books such as *Bumblebees Can't Fly*, a self-help book about strategies for remaining productive.

The buzzing wingbeats of a bumblebee's flight inspired the Russian composer Nikolai Rimsky-Korsakov to write the instrumental piece *Flight of the Bumblebee* around the year 1900. The frantic pace of the music – echoing the bumblebee's flight – has made it a standard showcase of a musician's skill. Perhaps because of this, the piece has been widely recorded in many diverse styles, from jazz to thrash metal. It can be heard in the films *The Green Hornet* and *Kill Bill*, amongst others, and a 1961 rock 'n' roll version ('Bumble Boogie' by B. Bumble and the Stingers) reached number 21 on the US charts.

Familiar hues

Bumblebees' familiarity and their distinctive, bright colour patterning has also resulted in them lending their name to a wide variety of other species: yellow-and-black striped orchids, shrimps, bats, frogs, beetles, fish and hummingbirds have all been named after bumblebees. Some characters (such as Bumblebee from *Transformers* and Bumblebee Man from *The Simpsons*) have gone a step further and added bumblebee's endearing clumsiness – their habit of bumbling around – to their yellow-and-black colour schemes.

Trains, gyrocopters, gliders and the world's smallest piloted aeroplane have all been named in tribute to

bumblebees, usually painted black and yellow to make the similarity more obvious. Bumblebees even frequently feature on honey jars, in place of the honeybees that actually make the honey. Presumably the endearingly bumbling fluffy, cuddly bumblebee is seen as a more attractive choice than the near-bald, angular honeybee.

Above: Bumblebees' ubiquity, iconic status and cuteness factor make them a regular feature in health food shop names and similar.

The intrinsic 'friendly' appearance of bumblebees can perhaps help explain their popularity. From the nursery we're raised to like and accept the fuzzy, friendly bumblebee, the star of a multitude of stories aimed at babies and children. It's only later, as childlike enthusiasm becomes replaced by the perceived coolness of not caring (often aided and abetted by a lack of accessible natural-history teaching or knowledge) that so many of us make the unwarranted transition to fearing bumblebees.

Bumblebees provide us with food and maintain the countryside in its current form by pollinating flowers in gardens, fields and hedgerows. Along the way they have been deified, vilified, misunderstood and ultimately recognised as the major wildlife *cause celebre* of the 21st century. Not bad for a 35-million-year-old group of hairy flower-visitors.

Glossary

Abdomen The third section of a bumblebee's body, containing the digestive and reproductive systems.

Antenna (plural antennae) Long, tubular sensory organs.

Anther The male, pollen-producing part of a flower.

Comb The aggregation of cells that make up a bumblebee nest.

Corbicula The 'pollen basket', an adaptation of the hind tibia to carry wetted pollen in bumblebees and honeybees.

Corolla The petals of a flower, usually when fused into a tube.

Cuckoo bumblebees Bumblebees that take over the nest of another species to rear their own offspring.

Drone Term for a male bee, usually used for honeybees.

Exoskeleton The external skeleton of many invertebrates, consisting of rigid plates joined by flexible cuticles.

Hibernaculum A small chamber in which the queen will hibernate.

Hibernation The state of becoming dormant to sleep through the winter.

Honey stomach A chamber at the front of the bee's abdomen to carry nectar back to the nest.

Inbreeding Breeding between closely related individuals, usually detrimental through accumulation of genetic issues.

Larva The mobile, eating-based juvenile stage of an insect, such as a caterpillar.

Mandibles The bumblebee's jaws.

Nectar A sugary fluid produced by plants as a reward for insects, mostly pollinators.

Nectar guides Ultraviolet markings on flowers pointing to the location of pollen or nectar.

Nectar-robbing Accessing the flower's nectar without moving pollen, usually by biting through the back of the flower.

Nectary A nectar-producing gland, usually at the base of the flower.

Ocellus (plural ocelli) Small light-sensitive structures (primitive eyes).

Ovipositor The egg-laying tube of most female insects. In bees it has been modified into a sting.

Parasitoid An organism that lives in or on another, and kills its host as it develops.

Pheromones Chemical messages used to inform or control other individuals.

Pistil The female part of the flower.

Pollen A plant's male sex cells, generally dispersed by wind or by insects.

Pupa The hard-shelled immobile life-stage of most insects, as they change from a larva into an adult.

Queen Reproductive female bumblebee or honeybee.

Scopa The 'pollen brush', a group of long hairs used to carry dry pollen in solitary bees.

Social bees Bees that build and live in communal nests, and have a worker caste.

Solitary bee A species in which every female makes her own nest, with no separate worker caste.

Stigma The receptive part of the female part of the flower, which receives pollen. Found on top of the ovary.

Style The 'stalk' between the ovary and the stigma.

Tarsus (plural tarsi) The bumblebee's five sectioned foot.

Thorax The second section of a bumblebee's body, mainly full of muscles and concerned with movement.

Tibia The bumblebee's 'shin', the section of leg just above the tarsus.

Worker A sterile female bumblebee or honeybee.

Further Reading and Resources

The Bumblebee Conservation Trust is Britain's only charity solely concerned with reversing the plight of bumblebees. Its website (www.bumblebeeconservation.org) has a range of information on managing your land or garden for bumblebees, identifying bumblebee species, and more. It also runs the UK's only standardised bumblebee abundance survey, BeeWalk, at www.beewalk.org.uk

The Bees, Wasps and Ants Recording Society (BWARS) is the national recording body for Hymenoptera, including bumblebees. Its website (www.bwars.com) has a range of resources including maps and species accounts. Its Facebook group, 'UK Bees, Wasps and Ants', is particularly excellent for getting identification help. Casual records – such as the bumblebees from your garden – should be sent to them at www.brc.ac.uk/iRecord

The Natural History Museum has a great research section on bumblebees, both British and from around the world, at www.nhm.ac.uk/research-curation/research/projects/bombus

Books

The past decade has seen a flurry of excellent books on bumblebees released. The 'bible' is Ted Benton's *Bumblebees* in the New Naturalist series (Collins, 2006). Unfortunately, as New Naturalists have become collectible, secondhand copies increasingly go to dealers for silly money, rather than to naturalists who would use them for the purpose they were written.

The first identification guide to all British bees in over a century was published in 2015 (*Field Guide to the Bees of Great Britain and Ireland*, Steven Falk and Richard Lewington, Bloomsbury Publishing) and contains detailed accounts of all the British bumblebees, including the extinct species.

Also excellent, both for identification and more general bumblebee ecology, is Prys-Jones and Corbet's *Bumblebees* in the Naturalists' Handbook series (Pelagic Publishing, 2011). In the field, my favourite is the *Field Guide to Bumblebees of Britain and Ireland* by Mike Edwards and Martin Jenner (Ocelli, 2009).

The first real bumblebee monograph – *The Humblebee, Its Life-history and How to Domesticate It* – was written by Frederick Sladen in 1912 and is still well worth reading. An out-of-copyright version can be downloaded from the Biodiversity Heritage Library at www.biodiversitylibrary.org/ia/humblebeeitslife00slad

Plants

British Wild Flower Plants has a very wide range of plants for mini meadows, etc. www.wildflowers.co.uk

Really Wild Flowers has plants, bulbs and seeds of native wildflowers, including shrubs and hedge trees. www.reallywildflowers.co.uk

Emorsgate Seeds sells British-sourced seeds of native wildflowers. wildseed.co.uk

Seeds Direct has a range of seed mixes for bees, and for different habitat and soil types. www.seeds-direct.org

Flora Locale has a list of native seed and wild plant suppliers. www.floralocale.org

Bumblebee species in Britain

Cryptic Bumblebee, *B. cryptarum*
(AKA: Cryptic White-tailed Bumblebee)
Great Yellow Bumblebee, *B. distinguendus*
Garden Bumblebee, *B. hortorum*
Brown-banded Carder Bumblebee, *B. humilis*
Tree Bumblebee, *B. hypnorum*
Heath Bumblebee, *B. jonellus*
Red-tailed Bumblebee, *B. lapidarius*
(AKA: Stone Bumblebee)
White-tailed Bumblebee, *B. lucorum*
(AKA: Small Earth Bumblebee)
Northern White-tailed Bumblebee, *B. magnus*
Bilberry Bumblebee, *B. monticola*
(AKA: Blaeberry Bumblebee, Mountain Bumblebee)
Moss Carder Bumblebee, *B. muscorum*
(AKA: Large Carder Bumblebee)
Common Carder Bumblebee, *B. pascuorum*
Early Bumblebee, *B. pratorum*
(AKA: Early-nesting Bumblebee)
Red-shanked Carder Bumblebee, *B. ruderarius*
Ruderal Bumblebee, *B. ruderatus*
(AKA: Large Garden Bumblebee)
Broken-belted Bumblebee, *B. soroeensis*
(AKA: Ilfracombe Bumblebee)
Short-haired Bumblebee, *B. subterraneus*
(last seen 1988, being reintroduced)
Shrill Carder Bumblebee, *B. sylvarum*
Buff-tailed Bumblebee, *B. terrestris*
(AKA: Large Earth Bumblebee)
Barbut's Cuckoo Bumblebee, *B. barbutellus*
Gypsy Cuckoo Bumblebee, *B. bohemicus*
Field Cuckoo Bumblebee, *B. campestris*
Red-tailed Cuckoo Bumblebee, *B. rupestris*
(AKA: Hill Cuckoo Bumblebee)
Forest Cuckoo Bumblebee, *B. sylvestris*
(AKA: Four-coloured Cuckoo Bumblebee)
Southern Cuckoo Bumblebee, *B. vestalis*
(AKA: Vestal Cuckoo Bumblebee)

Extinct in Britain
Cullum's Bumblebee, *B. cullumanus*
(last seen 1941)
Apple Bumblebee, *B. pomorum*
(last seen 1864)

Bumblebee-watching equipment

Watkins and Doncaster stock pretty much everything you need to catch or examine bumblebees – butterfly nets, collection pots and much more. www.watdon.co.uk

Anglian Lepidopterist Supplies carry a wide range of bumblebee-catching gear, including nets, collecting pots, and close-focus binoculars. www.angleps.com

Acknowledgements

This book would not have been possible without my partner Kate, who put up with me regularly disappearing to the study to write even as we tried to move house. She also took several of the pictures I needed, and provided much useful feedback on when I might be getting that bit too detailed on bumblebee minutiae!

Thanks must go to Julie Bailey and Alice Ward at Bloomsbury for offering me the chance to write a book in the first place, and for their support and help throughout the process. Last (but not least), my friends and colleagues (past and present) at the Bumblebee Conservation Trust, fighting to save British bumblebees and in the process creating an atmosphere where it's impossible not to fall for the furry little bees.

Image credits

Bloomsbury Publishing would like to thank the following for providing photographs and for permission to reproduce copyright material.

While every effort has been made to trace and acknowledge all copyright holders, we would like to apologise for any errors or omissions and invite readers to inform us so that corrections can be made in any future editions of the book.

Key t = top; l = left; r= right; tl = top left; tcl = top centre left; tc = top centre; tcr = top centre right; tr = top right; cl = centre left; c = centre; cr = centre right; b = bottom; bl = bottom left; bcl = bottom centre left; bc = bottom centre; bcr = bottom centre right; br = bottom right

AL = Alamy; FL= FLPA; G = Getty Images; NPL = Nature Picture Library; RS = RSPB Images; SS = Shutterstock

Front cover t Jacky Parker Photography/G, b Tony Sweet/G; **back cover** t Michael Durham/ Minden Pictures/FL, b Richard Becker/FL; **1** SS; **3** SS; **4** mtking/G; **5** Wayne Lynch/G; **6** Nick Upton/2020VISION/NPL; **7** t SS, b Andy Pay; **8** Michael Durham/Minden Pictures/FL; **9** t Konrad Wothe/Minden Pictures/FL, b Phil Savoie/NPL; **10** I love nature/G; **11** t Lizzie Harper, b Albert Lleal/Minden Pictures/FL; **12** Richard Comont; **13** SS; **14** SS; **15** Andy Sands/NPL; **16** Steven Falk/OUMNH; **17** t Louise Murray/ robertharding/G, b Anna Henly/G; **18** t SS, b Andrew Harrington/NPL; **19** Richard Comont; **20** Kate Ashbrook; **21** t Richard Comont, b Ammonite/NPL; **22** t Paul Hobson, b SS; **23** t Hans Christoph Kappel/NPL, b Piotr Naskrecki/ Minden Pictures/FL; **24** Alex Hyde/NPL; **25** SS; **26** Wild Wonders of Europe/Geslin/NPL; **27** t Regis Cavignaux/Biosphoto/FL, b Photo Researchers/FL; **28** SS; **29** Moritz Haisch/ EyeEm/G; **30** Ray Wilson/AL; **31** Kate Ashbrook; **32** Naturepix/AL; **33** Sue Robinson/AL; **34** Kate Ashbrook; **35** t Jonathan Lewis/G, b Visuals Unlimited, Inc./Robert Pickett/G; **36** Paul Hobson/FL; **37** Scott Camazine/AL; **38** t Christina Bollen/G, b David Boag/AL; **39** Richard Comont; **40** Richard Becker/FL; **41** blickwinkel/AL; **42** Richard Becker/FL; **43** Michael Durham/Minden Pictures/FL; **44** SS; **45** John B Free; **46** SS; **47** Albert de Wilde/Minden Pictures/FL; **48** Horst Sollinger/FL; **49** G E Hyde/FL; **50** Paul Starosta/G; **51** Kim Taylor/NPL; **52** SS; **53** t Peter Entwistle/ FL, b Dave Pressland/FL; **54** t Richard Becker/FL, b Sue Kennedy/RS; **55** Martin Gabriel/NPL; **56** Roger Tidman/RS; **57** t Gerry Ellis/FL, b Richard Comont; **58** Phil Cutt/RS; **59** t Richard Becker/AL, b SS; **60** SS; **61** Stephen Dalton/NPL; **62** t Steve Hopkin/G, b Westend61/G; **63** t Richard Becker/ FL, b IMAGEBROKER,ANDRE SKONIECZNY/FL; **64** t Westend61/G, b Jan Van Arkel/FL; **65** SS; **66** mikroman6/G; **67** AFP/Stringer/G; **68** Photo Researchers/FL; **69** t Friedrich (Klimpi) Loosli (Klimperator) / EyeEm/G, b Richard Newstead/G; **70** t Rod Teasdale, b Wally Eberhart/G; **71** t Pallab Seth/G, bl Phil Savoie/NPL, br Norbert Wu/FL; **72** t Oli Scarff/G, b SS; **73** t Nigel Cattlin/ FL, b Gary W. Carter/G; **74** WildPictures/AL; **75** Richard Becker/FL; **76** Bob Gibbons/FL; **77** t James Lowen/FL, b SS; **78** John Waters; **79** SS; **80** SS; **81** t Dave Pressland/FL, b Paul Hobson/NPL; **82** t Kerstin Hinze/NPL, b SS; **83** t Bernard Castelein/NPL, b David Tipling/NPL; **84** Wild Pictures/AL; **85** t Martin Siepmann/FL, b SS; **86** t Roger Tidman, b Mark Moffett/FL; **87** MD Kern/ Palo Alto JR Museum/NPL; **88** David Kjaer/RS; **89** Kate Ashbrook; **90** t Bob Gibbons/FL, b John Eveson/FL; **91** SS; **92** Phil Savoie; **93** ullstein bild/ Contributor/G; **94** t Cisca Castelijns, NiS/FL, b SS; **95** Nigel Cattlin/FL; **96** Reinhard Hölzl/ Imagebroker/FL; **97** t Francois Merlet/FL, b Richard Becker/FL; **98** Ernie Janes/RS; **99** tl Ben Philips/FL, tr Ben Philips/FL, b SS; **100** t Hocken Library, b ullstein bild/Contributor; **101** t SS, b MichaelGrantWildlife/AL; **102** Richard Becker/ FL; **103** Nigel Cattlin/FL; **104** David Hosking/FL; **105** bl SS, br Ernie Janes/RS; **106** Richard Becker/ FL; **107** t Nick Upton/RS, b SS; **108** SS; **109** t Bob Gibbons/FL, b Paul Hobson/FL; **110** SS; **111** Nick Upton/2020VISION/NPL; **112** t Nick Upton/NPL, b SS; **113** t Michel Gunther/FL, b SS; **114** Mike Lane; **115** t Bert Pijs/Minden Pictures/FL, bl Nick Upton/NPL, br Westend61 GmbH/AL; **116** David Woodgall/NPL; **117** MightyIsland/G; **118** Oramstock/AL; **119** WorldPhotos/AL; **120** Jastrow; **121** t Werner Forman/G, b The Museum of Witchcraft & Magic; **122** David Livingston/G; **123** Cath Harries/AL.

Index